P9-APB-727

METHODOLOGICAL ISSUES FOR HEALTH CARE SURVEYS

STATISTICS: Textbooks and Monographs

A SERIES EDITED BY

Vol. 1: The Generalized Jacknife Statistic, *H. L. Gray and W. R. Schucany*
Vol. 2: Multivariate Analysis, *Anant M. Kshirsagar*
Vol. 3: Statistics and Society, *Walter T. Federer*
Vol. 4: Multivariate Analysis: A Selected and Abstracted Bibliography, 1957-1972, *Kocherlakota Subrahmaniam and Kathleen Subrahmaniam* (out of print)
Vol. 5: Design of Experiments: A Realistic Approach, *Virgil L. Anderson and Robert A. McLean*
Vol. 6: Statistical and Mathematical Aspects of Pollution Problems, *John W. Pratt*
Vol. 7: Introduction to Probability and Statistics (in two parts), Part I: Probability; Part II: Statistics, *Narayan C. Giri*
Vol. 8: Statistical Theory of the Analysis of Experimental Designs, *J. Ogawa*
Vol. 9: Statistical Techniques in Simulation (in two parts), *Jack P. C. Kleijnen*
Vol. 10: Data Quality Control and Editing, *Joseph I. Naus* (out of print)
Vol. 11: Cost of Living Index Numbers: Practice, Precision, and Theory, *Kali S. Banerjee*
Vol. 12: Weighing Designs: For Chemistry, Medicine, Economics, Operations Research, Statistics, *Kali S. Banerjee*
Vol. 13: The Search for Oil: Some Statistical Methods and Techniques, *edited by D. B. Owen*
Vol. 14: Sample Size Choice: Charts for Experiments with Linear Models, *Robert E. Odeh and Martin Fox*
Vol. 15: Statistical Methods for Engineers and Scientists, *Robert M. Bethea, Benjamin S. Duran, and Thomas L. Boullion*
Vol. 16: Statistical Quality Control Methods, *Irving W. Burr*
Vol. 17: On the History of Statistics and Probability, *edited by D. B. Owen*
Vol. 18: Econometrics, *Peter Schmidt*
Vol. 19: Sufficient Statistics: Selected Contributions, *Vasant S. Huzurbazar (edited by Anant M. Kshirsagar)*
Vol. 20: Handbook of Statistical Distributions, *Jagdish K. Patel, C. H. Kapadia, and D. B. Owen*
Vol. 21: Case Studies in Sample Design, *A. C. Rosander*

Vol. 22: Pocket Book of Statistical Tables, *compiled by R. E. Odeh, D. B. Owen, Z. W. Birnbaum, and L. Fisher*
Vol. 23: The Information in Contingency Tables, *D. V. Gokhale and Solomon Kullback*
Vol. 24: Statistical Analysis of Reliability and Life-Testing Models: Theory and Methods, *Lee J. Bain*
Vol. 25: Elementary Statistical Quality Control, *Irving W. Burr*
Vol. 26: An Introduction to Probability and Statistics Using BASIC, *Richard A. Groeneveld*
Vol. 27: Basic Applied Statistics, *B. L. Raktoe and J. J. Hubert*
Vol. 28: A Primer in Probability, *Kathleen Subrahmaniam*
Vol. 29: Random Processes: A First Look, *R. Syski*
Vol. 30: Regression Methods: A Tool for Data Analysis, *Rudolf J. Freund and Paul D. Minton*
Vol. 31: Randomization Tests, *Eugene S. Edgington*
Vol. 32: Tables for Normal Tolerance Limits, Sampling Plans, and Screening, *Robert E. Odeh and D. B. Owen*
Vol. 33: Statistical Computing, *William J. Kennedy, Jr. and James E. Gentle*
Vol. 34: Regression Analysis and Its Application: A Data-Oriented Approach, *Richard F. Gunst and Robert L. Mason*
Vol. 35: Scientific Strategies to Save Your Life, *I. D. J. Bross*
Vol. 36: Statistics in the Pharmaceutical Industry, *edited by C. Ralph Buncher and Jia-Yeong Tsay*
Vol. 37: Sampling from a Finite Population, *J. Hajek*
Vol. 38: Statistical Modeling Techniques, *S. S. Shapiro*
Vol. 39: Statistical Theory and Inference in Research, *T. A. Bancroft and C.-P. Han*
Vol. 40: Handbook of the Normal Distribution, *Jagdish K. Patel and Campbell B. Read*
Vol. 41: Recent Advances in Regression Methods, *Hrishikesh D. Vinod and Aman Ullah*
Vol. 42: Acceptance Sampling in Quality Control, *Edward G. Schilling*
Vol. 43: The Randomized Clinical Trial and Therapeutic Decisions, *edited by Niels Tygstrup, John M. Lachin, and Erik Juhl*
Vol. 44: Regression Analysis of Survival Data in Cancer Chemotherapy, *Walter H. Carter, Jr., Galen L. Wampler, and Donald M. Stablein*
Vol. 45: A Course in Linear Models, *Anant M. Kshirsagar*
Vol. 46: Clinical Trials: Issues and Approaches, *edited by Stanley H. Shapiro and Thomas H. Louis*
Vol. 47: Statistical Analysis of DNA Sequence Data, *edited by B. S. Weir*
Vol. 48: Nonlinear Regression Modeling: A Unified Practical Approach, *David A. Ratkowsky*
Vol. 49: Attribute Sampling Plans, Tables of Tests and Confidence Limits for Proportions, *Robert E. Odeh and D. B. Owen*
Vol. 50: Experimental Design, Statistical Models, and Genetic Statistics, *edited by Klaus Hinkelmann*
Vol. 51: Statistical Methods for Cancer Studies, *edited by Richard G. Cornell*
Vol. 52: Practical Statistical Sampling for Auditors, *Arthur J. Wilburn*
Vol. 53: Statistical Signal Processing, *edited by Edward J. Wegman and James G. Smith*
Vol. 54: Self-Organizing Methods in Modeling: GMDH Type Algorithms, *edited by Stanley J. Farlow*
Vol. 55: Applied Factorial and Fractional Designs, *Robert A. McLean and Virgil L. Anderson*
Vol. 56: Design of Experiments: Ranking and Selection, *edited by Thomas J. Santner and Ajit C. Tamhane*
Vol. 57: Statistical Methods for Engineers and Scientists. Second Edition, Revised and Expanded, *Robert M. Bethea, Benjamin S. Duran, and Thomas L. Boullion*

Vol. 58: Ensemble Modeling: Inference from Small-Scale Properties to Large-Scale Systems, *Alan E. Gelfand and Crayton C. Walker*

Vol. 59: Computer Modeling for Business and Industry, *Bruce L. Bowerman and Richard T. O'Connell*

Vol. 60: Bayesian Analysis of Linear Models, *Lyle D. Broemeling*

Vol. 61: Methodological Issues for Health Care Surveys, *Brenda Cox and Steven Cohen*

Vol. 62: Applied Regression Analysis and Experimental Design, *Richard J. Brook and Gregory C. Arnold*

OTHER VOLUMES IN PREPARATION

METHODOLOGICAL ISSUES FOR HEALTH CARE SURVEYS

Brenda G. Cox
Research Triangle Institute
Research Triangle Park, North Carolina

Steven B. Cohen
National Center for Health Services Research
Rockville, Maryland

MARCEL DEKKER, INC. New York and Basel

Library of Congress Cataloging in Publication Data

Cox, Brenda G.
Methodological issues for health care surveys.

(Statistics, textbooks and monographs ; v. 61)
Includes index.
1. Medical care, Cost of--Longitudinal studies.
2. Medical care--Utilization--Longitudinal studies.
3. Health surveys--Methodology. 4. Medical care, Cost of
--United States--Longitudinal studies. 5. Medical care--
United States--Utilization--Longitudinal studies.
I. Cohen, Steven B., [date]. II. Title. III. Series.
[DNLM: 1. Data Collection--methods. 2. Health Surveys--
United States. WA 950 C877m]
RA410.5.C69 1985 362.1'028 85-4567
ISBN 0-8247-7323-3

MARCEL DEKKER, INC.
270 Madison Avenue, New York, New York 10016

Current printing (last digit):
10 9 8 7 6 5 4 3 2 1

PRINTED IN THE UNITED STATES OF AMERICA

Foreword

The demand for data addressing current problems and policy issues by legislators, government agencies, and administrators at federal, state, and local levels has grown steadily and substantially over the past 25 years. The growth in ability to consume larger and larger volumes of data has paralleled the spread of the computer into all aspects of the business of government. A major source of these data is the sample survey, by which information is gathered in personal or telephone interviews, or using a mail questionnaire, from samples of individuals, households, establishments and institutions, with the mode of collection and the sample depending upon the problem being addressed. This book represents a distinct and important contribution to the development of the sample survey method in general, and, in particular, to the application of the methodology of sample surveys to the collection of data on the utilization of health care services and the cost of those services to the public.

The explication of the methodological aspects of sample surveys in the context of medical care and expenditure data is singularly useful. Issues of health care cost containment and the delivery of quality health care have been major topics of concern to federal and state agencies for some time and can be expected to continue to challenge these agencies in the years ahead. The sample survey is a key tool essential to the capture of the information needed to address these issues and to derive cost effective solutions.

Longitudinal social surveys are receiving increased attention, motivated primarily by the growing demand for data that more accurately reflect the dynamic aspects of social phenomena. The 1977 National Medical Care Expenditure Survey (NMCES) and the 1980 National Medical Care Utilization and Expenditure Survey (NMCUES), both of which are discussed extensively in this book, were longitudinal surveys of families and individuals. Important methodological issues that arise in the design and conduct of longitudinal surveys and in the analysis of longitudinal data are identified by the authors. Useful approaches and solutions to these problems are presented.

This book is valuable in several other respects. First, it will be particularly valuable to those who will need to design and carry out field surveys of health care services and expenditures, that is, the practitioners. Second, it will be especially valuable to students since it not only provides technical discussion of all aspects of sample surveys, it does so in the context of application in a very real sense. The methodological problems and issues faced in the NMCES and NMCUES bring to life the technical aspects of the sample survey in a manner only rarely available in books heretofore. Third, it demonstrates the importance of technical expertise in the design and conduct of sample surveys, and the application of statistical theory and methods to control within acceptable limits the sampling and measurement errors which can occur in surveys.

It has been my pleasure to work with authors Brenda Cox and Steven Cohen on both the medical care surveys discussed in this book. I was very pleased with their decision to collaborate in writing it and I congratulate them on its completion. It should serve students and the sample survey practitioner community very well in the years ahead.

D. G. Horvitz, Ph.D.
Executive Vice President
Research Triangle Institute

Preface

This book is written primarily for health care professionals who wish to learn more about methodological issues associated with health care surveys. In view of the generality of the research described in many of the chapters, we believe the book will also be of interest to sampling statisticians and other survey researchers.

The material that is presented in this book was developed as a part of the authors' research activities for the 1977 National Medical Care Expenditure Survey (NMCES) and the 1980 National Medical Care Utilization and Expenditure Survey (NMCUES). Both surveys were longitudinal in design, collecting data on the health care experiences of a sample of individuals and their families during the data collection year.

The specialized nature of the two surveys is reflected by the research presented in this book. The authors do not claim to address all issues relevant to health care surveys. However, many of the problems encountered by NMCES and NMCUES - missing data for instance - are common to all sample surveys. For this reason, the authors believe the NMCES and NMCUES research will be of general interest.

In presenting this research, an attempt has been made to define first the methodological issue that is being addressed and then to describe the specific problem and solution for the NMCES or NMCUES study. Background material relating to sample design, weighting, missing data replacement, and data analysis is also presented.

Methodological Issues for Health Care Surveys is informally divided into three parts: sample design issues, missing and faulty data concerns, and analysis considerations. To lead into these discussions, Chapter 1 provides an overview of the use made of health care data and the type of data that are collected and references the major health data systems of the United States.

Chapter 2 provides background material for those readers unfamiliar with the techniques and terminology used in sample design. The chapter begins by discussing the importance of clarifying survey objectives and defining the target population. Then construction of the sampling frame and sample selection procedures are described.

Chapter 3 describes two alternate definitions for the observation unit of a longitudinal design. A study is then presented that compared the dwelling unit design to the household design for use in the 1980 NMCUES. Little difference in survey costs was found between the two designs. In addition, the dwelling unit design could not satisfy NMCUES data needs as well as the household design could. For these reasons, the recommendation was made to continue the use of a longitudinal household design to collect utilization and expenditure data.

Chapter 4 presents a study of data collection organization effect for the NMCES. NMCES was conducted by two survey organizations, each independently selecting a national sample and collecting survey data for that sample. The magnitude of the data collection organization effect was estimated for a wide range of demographic and health related variables and found to be in general negligible.

Chapter 5 presents a study of alternate, optimally allocated designs for a future NMCUES based upon linkage with the National Health Interview Survey (NHIS). These designs would select the sample for the NMCUES using the NHIS sample as the frame. When increased precision is needed for small domains, the study concluded that an optimally allocated design can produce the same precision at significant cost savings over an unlinked area sample design.

Chapter 6 compares household reports of the medical conditions associated with their visits to medical providers to the diagnosis reports obtained directly from the medical provider. The basic conclusion resulting from the study was that the household report does not appear to be an adequate predictor for the provider report. However, changes in the recoding system would likely produce greater agreement between the two reports.

Chapter 7 addresses the topic of weight development for survey data, including the use of weighting class and post-stratification adjustments for total nonresponse. These procedures are illustrated for the NMCES household survey.

Procedures that can be used to replace or impute missing data are described in Chapter 8. These techniques are illustrated in Chapter 9 using results from the NMCUES item imputation activities. In general, the NMCUES response rate tended to be high for socio-demographic variables (95 percent or more) and lower than would be ideally desired for income and charge data items.

Chapters 10 and 11 both address the topic of techniques to compensate for partial annual data in longitudinal surveys. Chapter 10 describes the results of a NMCES methodological investigation comparing alternate weighting and imputation techniques for partial data compensation. Based upon this study, the conclusion was reached that imputation was best but that weighting techniques were also appropriate. Chapter 11 describes the results of implementing attrition imputation for the NMCUES. Initial response rates and attrition rates over time are also presented.

A general discussion of statistical methods for analyzing complex survey data is presented in Chapter 12. Chapter 13 continues the topic of data analysis by describing methods to approximate the variance of survey estimates. The design effect model and relative variance curve methods are described. In a comparison made using NMCES data, the design effect model consistently yielded variance approximations superior in accuracy to those of the relative variance curve.

Chapter 14 describes the dynamic approach used in the NMCES for family-level analyses. In addition, families were classified according to whether their membership changed during the year.

Differences were observed between the two family groups in age, marital status, sex, and employment status of the family head.

Finally, Chapter 15 presents synthetic estimation procedures that obtain model-based estimates for areas too small to yield reliable survey statistics. Several procedures are described and empirically compared using data from the NMCES. The composite estimate consistently produced a better estimate than the sample-regression estimate or the post-stratified estimate.

Brenda G. Cox
Steven B. Cohen

Acknowledgments

The research described in this book was completed with the advice and support of staff from several government agencies and contract research organizations. The authors gratefully acknowledge this assistance in preparing the book.

Dr. Cohen conducted the research described in this book as an employee of the National Center for Health Services Research (NCHSR) using data from the National Medical Care Expenditure Survey (NMCES) that NCHSR sponsored. The chapters written by Dr. Cohen (1,4,7,10, 12 to 15) are in the public domain and are not subject to copyright.

The authors give special thanks to the staff of NCHSR for their assistance in compiling the research described in this book. Dr. Daniel G. Walden, Senior Research Manager, provided continued support, inspiration, and guidance throughout the book's development and a thorough and insightful review of the manuscript. In his capacity as Acting Director, Dr. Donald G. Goldstone was particularly supportive of the research effort at a critical time in its development. Dr. John Marshall, Director, and Dr. Samuel P. Korper, Director of Intramural Research, also provided support and encouragement. Dr. Marc Berk, Ms. Vicki Burt, Dr. Judith Kasper, Mr. Samuel Meyers, Dr. Louis Rossiter, and Dr. Gail Wilensky provided helpful suggestions and reviews of selected chapters. Programming support for Dr. Cohen's analyses was provided by Ms. Burt and by Social and Scientific Systems, Inc. of Bethesda, Maryland.

Dr. Cox conducted much of the research described in this book as an employee of the Research Triangle Institute (RTI) under three contracts RTI had with the Federal Government. Assisted by its subcontractors, the National Opinion Research Center (NORC) and Abt Associates Inc., RTI conducted the 1977 NMCES under Contract No. HRA-230-76-0268 from NCHSR, with Dr. Daniel Walden as the Project Officer. Methodological research needed to plan the 1980 NMCUES was conducted under Contract No. 233-78-2101 from the National Center for Health Statistics (NCHS) and the Health Care Financing Administration (HCFA) with Mr. Earl Bryant and Dr. Allen Dobson as the Project Officers. With the assistance of NORC and SysteMetrics, Inc. as subcontractors, RTI conducted the 1980 NMCUES under Contract No. HRA-233-79-2032 from NCHS and HCFA, with Mr. Robert Fuchsberg and Dr. Allen Dobson as Project Officers.

The authors gratefully acknowledge reviews and assistance provided by Mr. Bryant, Dr. Dobson, Mr. Fuchsberg, and Dr. Walden and their cooperation in obtaining the required contract releases. In addition, the authors would also like to thank Mr. Robert Wright of NCHS who provided background materials for the book as well as advice based upon his review of selected chapters. The authors also appreciate the assistance of Dr. Herbert Silverman of HCFA in expediting the review and release of the book.

The authors would also like to express their appreciation to the Research Triangle Institute (RTI) and its staff for their encouragement in completing the book. Part of Dr. Cox's effort in preparing this book was underwritten by RTI who awarded her a Professional Development Award in 1982. Dr. Daniel G. Horvitz, Executive Vice President of RTI and Project Director for NMCES and NMCUES, was a constant source of inspiration and insight during the projects themselves and in completing this book. The authors are also indebted to Dr. Ralph Folsom for his willingness to provide technical advice and consultation as needed for the book. Dr. Folsom, Associate Director of the Center for Survey Statistics, served as Project Director for the NMCES Methods and Analysis Contract and as Sampling Task Leader for the NMCES and NMCUES studies. Other RTI staff who made this book possible by their support and encouragement include Dr. James Chromy, Vice President of Survey and Computing Sciences; Dr. Robert Mason, Director of the Center for Survey Statistics; and Dr. Judith Lessler, Department Manager in the Center for Survey Statistics. Critical to the preparation of this book was the typing support provided by Ms. Brenda Porter who patiently completed revision after revision of the text with minimal complaints.

Finally, the authors extend their heartfelt thanks to family and friends for patience in listening to endless talk about the book and for reassurances when we needed them. Particular thanks are extended to Paula and Celia Cohen whose sacrifice of time made the difference.

Contents

METHODOLOGICAL ISSUES FOR HEALTH CARE SURVEYS

1
Measuring Health Care Parameters in Sample Surveys

The American health care system has traditionally concerned itself with the provision of services to insure the maintenance of health, the prevention and diagnosis of disease, and treatment and rehabilitation. The health care needs of the population are met through the delivery of services by medical care providers and health care facilities, with additional health maintenance and preventive care instructions provided by educational programs. The financial mechanisms that support it are inseparably wed to the system in their stimulative effect on system development and the limitations they present to the system's expansion and functional capacities.

The identification of health care needs, the manpower required to meet these needs, and their appropriate distribution are components of the system's function. Related features include the investment in biomedical and epidemiological research, the education of health care providers, and the development of new medical technologies to broaden and improve the quality of care. The ability of the health care system to respond to the dynamic needs of the nation and the adequacy of that response is governed by the availability of relevant data for planning, and its use to establish baselines and assess change. Data are generally collected on the users of the system as well as the institutions and providers of care. Sociodemographic, economic, medical utilization, expenditures, insurance coverage, health status, and diagnostic measures are among the many health care indices of the user population. These measures are used in the evaluation process and as an aid to health planning and health services research. For effective administration of health services, information is collected on manpower and facility requirements, access to care, satisfaction with care, manpower shortage areas, and financing constraints.

The more visible impacts on the assessment of the nation's health and the identification of deficiencies in the health care system have usually been made by federally sponsored surveys. These surveys have served to stimulate the formulation of health policy and legislation to ameliorate identified limitations. Usually, these surveys are conducted directly by an agency of the federal government or funded by the agency. Often, an agency will fund a national health care survey and assume responsibility for questionnaire design and data analyses, but contract for the collection of data by an independent organization.

Many prominent national health care surveys have been conducted and funded by sources other than the federal government, however. These include:

- the National Survey of Access to Medical Care, conducted in 1976 by the Center for Health Administration Studies of the University of Chicago and primarily supported by the Robert Wood Johnson Foundation;
- the Periodic Survey of Physicians, conducted annually since 1966 by the American Medical Association to determine practice characteristics, patient profiles, and physician income;
- the Annual Survey of Hospitals, conducted annually since 1961 by the American Hospital Association to obtain data on hospital admissions, assets, average daily census of patients, revenue, and occupancy; and
- the Survey of Dental Practice, conducted every two or three years since 1950 by the American Dental Association to determine practice characteristics, expenses, income, and standard demographic profiles of dentists.

Numerous health surveys have also been conducted by state and local governments. Their focus is generally quite specific in nature (i.e., success of community mental health centers, prevalence of drug abuse in specified urban areas, maldistribution of physician supply in select rural settings), although their implications may be national in scope.

1.1 The Uses of Health Care Data

Historically, there has been a high degree of diversity in the use of health care data. This diversity stems from the varying goals and needs of user populations, which are affected by the political, economic and social conditions that prevail at any given time. Health concerns have shifted from infectious diseases and epidemics to preventive medicine and human ecology, from contagious diseases of childhood to chronic degenerative conditions associated with aging, from national health insurance to financing that places a greater burden on the patient, and from cost containment legislation for regulation of the health care system to consideration of the benefits of a free market system. The most prominent users of health care data are federal, state, and local governments, private industry, foundations, educational institutions, and concerned individuals. Several general, though by no means exhaustive, uses of health care data include:

- program planning and evaluation;
- identification of the medically underserved and others with limited access to health care delivery systems;
- health education and health promotion;
- epidemiological, biomedical and health services research;
- market research;
- measurement of the impact and extent of illness;
- measurement of the use of health care services, related medical expenditures, and sources of payment for care; and
- identification of public health problems.

The common thread that binds these varied topics is the use of health care data to establish a baseline from which to assess change.

1.2 The Types of Health Care Data Collected

Health care surveys are primarily designed to study three distinct groups: the general population, health care providers, and health care facilities. Although there is an interdependent relationship between these components within the framework of the health care system, each population generates a unique set of survey parameters.

National data on the incidence of acute illness and accidental injuries, the prevalence of chronic conditions and impairments, the extent of disability, and the utilization of health care services are obtained through the National Health Interview Survey (NHIS). Conducted annually by the National Center for Health Statistics (NCHS), the NHIS is a cross-sectional survey of approximately 40,000 households selected to represent the civilian, noninstitutionalized population of the United States (U.S.). The sample data measure demographic and socio-economic characteristics, health status, the impact and extent of illness, and the use of health care services. Supplements to the basic questionnaire are used that focus on special health topics in response to current interests. These topics have included immunizations, smoking, home health care, eye care, residential mobility, retirement income, and health insurance (National Center for Health Statistics, 1981, pp. 11-14).

Prevalence data for specific diseases and conditions of ill health are collected concurrently with measurements of the nutritional status of the American population in the National Health and Nutrition Examination Survey, also conducted by the National Center for Health Statistics. The 1976-1980 survey represented the U.S. population between the ages of 6 months to 74 years, with approximately 28,000 individuals completing the nutrition component and 21,000 undergoing examination. The data collection team consisted of specially trained interviewers, in addition to physicians, nurses, dentists, dieticians, and medical laboratory and x-ray technicians who conducted direct physical examinations and laboratory tests. The nutrition measures that were obtained included: (1) the measurement of dietary intake, (2) hemotological and biochemical tests (blood pressure, serum cholesterol levels), (3) body measurements, and (4) signs of high risk nutritional deficiency. In measuring prevalence levels of specific diseases and conditions, emphasis was placed on diabetes, kidney and liver functions, allergies, speech pathology, and heart disease (National Center for Health Statistics, 1981, pp. 16-19).

Another NCHS-sponsored survey, the National Ambulatory Medical Care Survey is a perennial source of statistical data on the ambulatory medical care provided by office-based physicians to the U.S. population. The target population consists of office-based visits to physicians who are not employees of the federal government and who are engaged in the provision of direct care to ambulatory patients. Approximately 3,000 physicians are included in the sample to provide information concerning patient visits, date and duration of visit, and patient characteristics including age, race, sex, ethnicity, and reason for visit. Additional information is collected on whether previous physician contacts were made for the same problem, the length of time since problem onset, diagnostic and therapeutic services provided, and disposition. General information is also obtained on the etiology and epidemiology of selected health conditions. Survey data permit research on the use, organization, and delivery of ambulatory care (National Center for Health Statistics, 1981, pp. 21-23).

Four other medical care provider surveys are of general interest to health care researchers. The American Medical Association has conducted the Periodic Survey of Physicians since 1966. Information is collected on practice characteristics, patient profiles, hours and weeks worked, professional income, professional expenses, professional fees and ancillary personnel data. Mathematica's Telephone Survey of Physician Capacity Utilization collected data in 1973 to 1975 on physician location, speciality status, board certification, practice characteristics, hours worked, waiting time, and fees for a routine visit. The American Medical Association conducted the Survey of 1960 and 1965 Graduates of American Medical Schools, which obtained information on physician residence prior to medical school, factors influencing physician choice of practice location, and previous practice locations. Finally, the Physician Master File of the American Medical Association provides information on every known physician and medical student in the United States and on American physicians practicing abroad. Two types of data items are included: (1) historical data on training, license, certification, and specialty choice; and (2) current professional activities, type of practice, and practice location (Langwell, et al, 1982).

With regard to the institutional component of the health care system, national statistics on inpatient utilization of short-stay hospitals are available from the National Hospital Discharge Survey. The sample is restricted to those hospitals with six or more beds and an average length of stay for all patients of less than 30 days. In 1979, 496 short-stay hospitals were selected from a target population of approximately 8,000 hospitals, yielding a sample of approximately 215,000 discharges. Information is obtained concerning the patient characteristics, length of stay, diagnoses, surgical operations, and use of care patterns for hospitals of differing size and ownership in each of the four major geographic regions of the United States (National Center for Health Statistics, 1981, pp. 20-21).

The NCHS National Nursing Home Survey provides continuing data on the most rapidly expanding sector of the health care industry. The 1977 survey obtained information on the characteristics of nursing homes, residents, staff, and services provided, from a sample of nursing homes selected to represent the nation. Data were also collected on facility costs in provision of care, certification for participation in Medicaid and Medicare programs, and the relationships that exist between utilization, services offered, charges for care, and the cost of providing care (National Center for Health Statistics, 1981, pp. 23-25).

Three other data sources provide national data for health care facilities. The National Master Facility Inventory is an annual census of facilities, providing data on ownership, the major type of service offered, the number of beds, patients, staffing patterns, and financing (National Center for Health Statistics, 1981, pp. 26 to 27). The American Hospital Association Annual Survey of Hospitals includes all registered hospitals in the United States. Information is obtained on hospital admissions, assets, beds, average daily census, expenses, revenue, occupancy, personnel, length of stay, and facility characteristics. The American Hospital Association also conducts the Survey of Hospital Service Charges, which compiles data for all hospitals on patient charges by services rendered (Langwell, et al, 1982).

1.3 National Health Care Utilization and Expenditure Surveys

The rapidly rising cost of medical services in the United States in recent years, together with a continuous effort to improve the quality, effectiveness, and availability of health care, has led to a continuing need for comprehensive data for individuals and families on health status, patterns of health care utilization, charges for services received, and payers and amount paid. The 1977 National Medical Care Expenditure Survey (NMCES) and the 1980 National Medical Care Utilization and Expenditure Survey (NMCUES) are two of a series of national health care surveys planned to provide these data on a regular basis. These surveys permit in-depth statistical descriptions of the utilization of health care services and the associated costs for various population segments, including the nation as a whole. They also provide valuable data for the evaluation of current public programs such as Medicare and Medicaid, for the assessment of inequity in access to the health care delivery system, and for the comparison of alternative solutions to health policy issues.

A comprehensive statistical description of the use of health care services and patterns of health expenditures and health insurance coverage in 1977 is provided by the National Medical Care Expenditure Survey (NMCES). The survey was designed and funded by the National Center for Health Services Research and cosponsored by the National Center for Health Statistics. To determine the utilization and financing mechanisms of health services in the United States in 1977, data were obtained in three complementary stages which surveyed: (1) approximately 14,000 randomly selected households in the civilian noninstitutionalized population, (2) physicians and health care facilities providing care to sample households during 1977, and (3) employers and insurance companies insuring the sample households (Cohen and Kalsbeek, 1981).

In the household interview portion of the NMCES, respondents were interviewed six times at approximately twelve-week intervals beginning in early 1977 and ending in mid-1978. Data were collected on use of medical services, charges for services and sources of payment, number and type of disability days, and medical conditions. Additional questions focused upon employment, health insurance coverage, access to care, barriers to care, ethnicity, income, and assets. For a 32 percent subsample of the household respondents, each of their medical providers was questioned on patient visits, diagnosis, charges, and sources of payment.

While the household survey focused on the demand for medical care, a component survey of physicians considered the supply of medical care. The Physicians' Practice Survey was established to provide data on the characteristics of direct patient care physicians providing care to the civilian noninstitutionalized population of the U.S. during 1977. Physicians were questioned regarding their place and mode of practice, hours worked weekly, the site of care, their age, speciality, board certification and practice, and other income.

The Health Insurance Employer Survey was a substudy of the NMCES that focused on the employers and insurance companies of the individuals included in the household survey. This survey component verified coverage information and obtained supplemental information with respect to health insurance benefits and premiums. Additional information for employers, regarding the number and characteristics of their employees during 1977, their total 1977 health insurance premiums, the number and salary level of employees eligible to participate in their health insurance plans, and total payroll expenditures, were obtained in the Employer Health Insurance Cost Survey component of the NMCES (Bonham and Corder, 1981).

The National Medical Care Utilization and Expenditure Survey (NMCUES) is the second survey in the series and has goals which build upon the experiences of the NMCES. This survey was sponsored by the National Center for Health Statistics in collaboration with the Health Care Financing Administration to provide detailed utilization and expenditure data for the civilian, noninstitutionalized population of the United States in 1980, with additional emphasis on Medicare and Medicaid recipients.

The National Household Survey (HHS) component of the NMCUES was based upon a stratified cluster sample of 7,200 dwelling units selected so as to represent the civilian noninstitutionalized population of the United States in 1980. Repeat interviews were conducted with the initial panel of 6,600 responding households at approximately twelve-week intervals beginning in early 1980 and ending in mid-1981. Data on health insurance coverage, episodes of illness, the number of bed days, restricted activity days, hospital admissions, physician and dental visits, other medical care encounters, and purchases of prescribed medicines were obtained in the household component. Additional information about health conditions, access to medical services, medical provider characteristics, services provided, charges, sources and amounts of payment was also collected. The survey allows for a determination of the persons using specific types of care and their costs and method of payment.

In addition to this household survey, NMCUES also included four State Medicaid Household Surveys (SMHS) of Medicaid households. From administrative record data provided by California, Michigan, New York, and Texas, a clustered list sample of Medicaid cases was selected from each state. The selection procedures yielded aid-category balanced samples of 1,000 cooperating Medicaid households per state. Using the same instrument and data collection procedures as the National Household Survey, 1980 health care data were collected for the four state samples of Medicaid recipients and their households, again using five data collection rounds approximately twelve weeks apart.

A third component, the Administrative Records Survey, extracted administrative record data from Medicaid and Medicare files. Medicaid eligibility in 1980 was collected for all persons reported as covered in the household surveys. In addition, Medicaid claims data were extracted for sample individuals in the California, Michigan, New York and Texas State Medicaid Household Surveys. For older persons in the HHS and SMSH, Medicare program eligibility was verified and charge and payment data extracted from the administrative Medicare files (National Center for Health Statistics, 1981, pp. 14 to 16).

1.4 Cross-Sectional and Longitudinal Survey Designs

National sample surveys are generally characterized by cross-sectional or longitudinal designs. Cross-sectional surveys provide a snapshot of population characteristics at a fixed point in time through a one-time survey of the population. An advantage of the cross-sectional design is the timeliness of the data. The processing time from data collection through input to computer and the derivation of estimates is greatly reduced when contrasted with the time requirements of longitudinal designs, which expand the data base with time series observations for each sampled individual. In addition, reduced costs are achieved through the administration of a one-time survey versus repeated interviews. Longitudinal surveys collect data on more than one occasion from the sample members of the population of interest in order to measure change and to obtain data for time periods too long to recall accurately in one interview. Longitudinal data are often essential for providing accurate descriptions of variations in population characteristics that are sensitive to changes in time. The advantages and disadvantages of the two designs are not absolute and must be balanced by the underlying aims of the proposed survey.

A major limitation of using cross-sectional designs for surveys such as the NMCES and NMCUES is the reference period that is needed, often spanning an entire year. Long reference periods result in more of both types of reporting errors, errors of omission and erroneous inclusion through forward telescoping. Errors of omission are characterized by the respondent forgetting an illness episode or expenditure or inaccurately recalling the event as happening outside the time period of interest (backward telescoping). For health care utilization surveys, these omissions are not random, but are usually concentrated among short term illnesses requiring no hospitalization and routine visits to physicians (Sudman and Lannom, 1979). With respect to forward telescoping, the episode is remembered in error in that the episode is viewed as occurring within the time period of interest when in fact it occurred earlier. In addition to reducing the reference period by using repeated interviews, errors due to telescoping decline and memory loss can be further minimized by interviewing techniques which include probing, submission of diaries to the respondent, and computer generated summaries. Summaries that describe the responses provided by the respondent in previous interviews of a panel design allow corrections for omissions and telescoping errors. Both the NMCES and NMCUES surveys made use of these techniques to minimize errors of omission. In addition, short recall periods of two to three months in duration were structured into the survey design to limit the potential for telescoping errors.

Longitudinal survey designs are also adopted to provide a mechanism to assess change in the behavior of a population over a specified time period. Often referred to as panel designs, they allow the measurement of seasonal variations in population characteristics. This capacity is of primary importance in major national health surveys, which attempt to measure the health status and morbidity levels of the population. Since these measures are sensitive to seasonal developments (i.e. climatic changes), a point estimate in time would have serious limitations, in its exposure to the risks of seasonal, secular, and catastrophic variation. Repeated interviews over an entire time period, usually a year in duration, may lead to better statistical inferences than a single one-time survey. The NMCES and NMCUES surveys were both structured as panel designs to provide data for analyses of health issues which include the extent of and reasons for changes in health insurance coverage over time, and the cost of illness for various diagnoses that are sensitive to seasonal variation.

Using NMCES data, Wilensky, et al (1981) estimate that the uninsured population for 1977 varied from a high of 25.9 million in the first quarter of 1977 to a low of 22.9 million in the second quarter. These figures are derived from two different populations experiencing lack of insurance coverage. Approximately 18.4 million were uninsured for all of 1977, whereas an additional 16.2 million were uninsured during part of 1977. National estimates of the uninsured population are also available from the National Health Interview Survey which has a cross-sectional design. According to the NHIS estimate for 1976, about 23.2 million individuals were without insurance coverage, although this estimate reflects the absence of private insurance at the time of the interview. The cross-sectional data does not allow for characterization of the part-time insured, a population group which affects a dynamic change in the composition of the uninsured population over the year. The capability of panel surveys to characterize populations that exhibit a change in composition is also illustrated in a study investigating shifts in Medicaid eligibility for 1977 (Wilensky et al, 1980).

In this text, we will concentrate on methodological issues that relate to health care surveys, particularly those with longitudinal designs and target populations consisting of the civilian noninstitutionalized population of the United States. We will be relying upon our experiences and investigations for two major national health care surveys: the 1977 National Medical Care Expenditure Survey and the 1980 National Medical Care Expenditure and Utilization Survey.

References

Bonham, G. S., and L. S. Corder (1981). NMCES Household Interview Instruments, NHCES Instruments and Procedures 1, DHHS Publication No. (PHS) 81-3280, Washington, DC: U.S. Government Printing Office.

Cohen, S. B., and W. D. Kalsbeek (1981). NMCES Estimation and Sampling Variance in the Household Survey, NHCES Instruments and Procedures 2, DHHS Publication No. (PHS) 81-3281, Washington, DC: U.S. Government Printing Office.

Langwell, K. M., J. D. Bobula, S. F. Moore, and B. B. Adams (1982). An Annotated Bibliography of Research on Competition in the Financing and Delivery of Health Services, NCHSR Report Series, DHHS Publication No. (PHS) 83-3326, Washington, DC: U.S. Government Printing Office.

National Center for Health Statistics, N. D. Pearce (1981). Data Systems of the National Center for Health Statistics. Vital and Health Statistics, Series 1, No. 16. DHHS Publication No. (PHS) 82-1318, Washington, DC: U.S. Government Printing Office.

Sudman, S. and L. B. Lannom (1979). A Comparison of Alternative Panel Procedures for Obtaining Health Data. Paper available from the Survey Research Laboratory, Urbana, Illinois: University of Illinois.

Wilensky, G. R., D. C. Walden, and J. A. Kasper (1980). The Changing Medicaid Population. Proceedings of the American Statistical Association, Section on Social Statistics, 165-170.

Wilensky, G. R., D. C. Walden, and J. A. Kasper (1981). The Uninsured and Their Use of Health Services. Proceedings of the American Statistical Association, Section on Social Statistics, 327-332.

2
Sample Design Considerations for Health Care Surveys

Health care surveys are of necessity based upon samples rather than complete censuses of the population of interest. Cost considerations typically lead the health care researcher to reject the idea of interviewing the entire population. In addition, timely production of survey statistics will usually dictate that the researcher restrict data collection to a sample of the total population. Using standard sampling methodology, the researcher can make statistical inferences with known precision about the population of interest. Further, sample data may even be more accurate than census data when the relatively small number of sample units facilitates the commitment of additional resources for personnel training, nonrespondent follow-up, and more accurate measurement methods such as physical examinations.

With careful attention to detail, the health care researcher can design a sample survey that satisfies the inferential needs of the study. This chapter discusses sample design considerations for health care surveys with illustrations from NMCES and NMCUES. Emphasis will be placed upon the effect that these sample design components have on the precision and accuracy of survey estimates. Accuracy refers to how close the sample estimate is to the parameter being estimated. Precision refers to the extent to which multiple measurements produce the same sample estimate. Accuracy is usually measured by the mean square error of the estimate and precision by the variance of the estimate (Kotz and Johnson, 1982, p. 15).

2.1 Determination of Survey Objectives

The first step in the design of any survey is to develop a detailed list of survey objectives. These objectives will motivate the entire sample design. The population to be sampled and the data to be collected are directly dependent upon the survey objectives. Often the mode of data collection is limited by the survey objectives because of the type of data to be collected. By clearly specifying the survey objectives in advance, the researcher is protected against the embarassing situation of not having adequate precision for survey estimates or of not having the capability to respond to important questions.

In developing the objectives, the planner should formulate as complete a list as possible. The objectives should focus on the uses to which the data are to be put and the questions that the data must answer. As an illustration of defining objectives, the Background Section of the NMCES report series documents (e.g., Cohen and Kalsbeek, 1981) provides the following implicit statement of the objectives of the National Medical Care Expenditure Survey.

> Analyzing how Americans use health care services and determining the patterns and character of health expenditures and health insurance are the goals of a landmark study being conducted by the National Center for Health Services Research (NCHSR). The study will provide important information and analyses on a number of issues:
> • The cost, utilization, and financing implications of various national health insurance proposals.
> • The influence of Medicare and Medicaid programs on the use of medical services and the costs of providing care.
> • The extent of and reasons for changes in Medicaid participation over time.
> • The extent to which different government programs at the Federal, state, and local levels affect access to care in different treatment settings.
> • The distribution of tax benefits to individuals and business under current tax laws concerning medical and health insurance expenses and the potential changes in the distribution of benefits if these laws were to be changed.
> • The costs of illnesses for various diagnoses in different treatment settings.
> • The breadth and depth of coverage and the proportion of medical costs paid by health insurance.

Another example of survey objectives is provided by the following excerpt from _Data Systems of the National Center for Health Statistics_ (National Center for Health Statistics, 1981, p. 14) concerning the National Medical Care Utilization and Expenditure Survey (NMCUES).

> NMCUES is designed to be directly responsive to the continuing need for statistical information on the health care expenditures associated with health services utilization for the entire U.S. population. Cycle I was designed and conducted in collaboration with the Health Care Financing Administration to provide detailed utilization and expenditure data for persons in the Medicare and Medicaid populations. NMCUES will produce estimates over time for evaluation of the impact of legislation and programs on health status, costs, utilization and illness-related behavior in the medical care delivery system.

The initial set of objectives are usually general statements of survey intent that must be specified in greater detail to develop the sample design and data collection plan. For instance, the precision requirements for survey estimates have to be evaluated and cost constraints determined. The survey objectives may necessitate a high level of response or the use of record check components. Subpopulations for which separate estimates are needed should be designated as well. These reporting groups are commonly referred to as domains. A final specific set of survey objectives should be formulated prior to data collection that details domains of interest and the required level of precision for domain estimates. Thereafter, the survey should be evaluated periodically in terms of how effectively it is satisfying these stated objectives.

2.2 Definition of the Target Population

The target population for a survey is the entire set of elements about which the survey data are to be used to make inferences. For health care surveys, the target population is usually composed of persons or providers of medical care. The definition of the target population for a particular survey follows directly from the survey objectives.

As an example, the NMCES target population was the civilian, noninstitutionalized residents of the United States in 1977 (Cohen and Kalsbeek, 1981). The reasoning leading to the restriction of the survey to civilians was that complete expenditure data were already available to the government for military personnel. The population was further restricted to residents of the United States since the health care experiences of nonresidents have little impact on public policy. The exclusion of institutionalized individuals had a different origin. The utilization and expenditures of institutionalized individuals do have an impact on public policy and data are not available for individuals in non-federal institutions. However, because of the difficulty of obtaining information from individuals in institutions, who may be physically or mentally incapacitated, it was decided that such individuals could only be included in a specially designed survey of individuals in institutions. Finally, the survey objectives required a high level of accuracy for the estimates of expenditures and utilization. This led to NMCES including component surveys such as the Medical Provider Survey, the Health Insurance/Employer Survey, and the Physicians' Practice Survey. Each of these component surveys had target populations defined and survey objectives developed. The Physicians' Practice Survey supplied data on the practice characteristics of direct care physicians. The Medical Provider Survey obtained expenditure and utilization data for the sample individuals in the household survey from their physicians and other medical care providers. The Health Insurance/Employer Survey verified insurance coverage and obtained copies of applicable policies.

Similar reasoning to that of NMCES led the NMCUES to be restricted to the civilian, noninstitutionalized residents of the United States in 1980. To facilitate more detailed analyses for the Medicaid population, surveys of households receiving Medicaid benefits were conducted in California, Michigan, New York, and Texas. The State Medicaid Household Surveys (SMHS) restricted the target population to these four states because of the difficulty and costs associated with obtaining and processing Medicaid case records for each of the 50 states. The four state programs studied in the SMHS accounted for 34 percent of all Medicaid recipients and 41 percent of all Medicaid payments in 1979 (Muse and Sawyer, 1982, p. 4). The target population for each of the state surveys was defined to be the universe of Medicaid households within the state. Since a study objective was to measure changes in Medicaid eligibility for households and individuals, Medicaid households were defined to be all families that contained at least one noninstitutionalized Medicaid beneficiary in November 1979. In addition to these household survey components, administrative records were examined to check eligibility and obtain claims data for sample persons reporting Medicaid and Medicare coverage (Cox, Piper, and Folsom, 1983).

How effectively a survey satisfies its stated objectives depends upon the extent to which the survey estimates agree with the equivalent parameters of the target population. The quality of the estimates is directly related to the adequacy of the procedures used in sample selection. The sample design should revolve around producing estimates with the most accurate representation of the target population within the available resources.

2.3 Use of Probability Sampling

Probability sampling refers to sampling in which a list or mechanism is developed for enumerating the target population (i.e., the sample frame) and every unit on the frame is given a known, nonzero probability of being selected for inclusion in the survey. Rosander (1977, pp. 4-5) lists the following ten advantages possessed by a probability sample that is properly designed and implemented:

1. It provides a quantitative measure of the extent of variation due to random effects.
2. It provides data of known quality.
3. It provides data in a timely fashion.
4. It provides acceptable quality data at minimum cost.
5. It has a built-in method of estimation.
6. It forces a sharper analysis of the problem and a clearer definition of terms.
7. It forces better control over nonsampling sources of error, that is, over bias.
8. Mathematical statistics and probability can be applied to analyze and interpret the data.
9. It can yield better quality data than a census or 100% tabulation.
10. It provides within itself a basis for progressive improvement.

To summarize, the overall advantage of probability sampling is that a defensible basis exists for inference from sample to population.

Probability sampling is unquestionably superior to subjective sampling in which personal judgements are made by survey planners or field interviewers as to which population elements are to be included in the survey. Even when attempts are made to judgmentally select "representative" units or to select units from all parts of the frame (i.e., quota sampling), the subjective sample does not possess the advantages of probability sampling and hence the survey results cannot be used to make statistical inferences about the target population.

The need to use valid probability sampling methods in survey sampling cannot be overemphasized. As early as 1960, Deming noted: (p. vi):

> A sampling plan to be satisfactory today, must not only produce a figure for an estimate of the value of an inventory, or of a certain class of account, or of the number of consumers that have some specific characteristic, or a measure of performance of a product in service, or of a medical treatment, to name a few examples; it must also

> provide demonstrable measures of the reliability of this estimate. The same thing is true of figures issued by a government agency on employment, unemployment, expenditures, production of lumber, retail sales by the month, housing conditions, etc.

The only exception to this rule made by modern survey statisticians is when small samples are drawn for use in preliminary pilot studies designed to test survey instrumentation and data collection procedures.

2.4 Construction of the Sampling Frame

The *sampling frame* for a survey is the list or mechanism used to enumerate population elements for sample selection purposes. The ideal frame is an actual list of population elements where every element is listed once and only once and where every listing represents an element of the target population. This ideal frame contains unique identifying information, address or location information, and auxiliary information about the element. As an example, the ideal frame for a national survey of hospitals would be a list of every hospital in the United States with name, address, and characteristics that related to research questions, such as number of beds and annual revenues. No hospital would be listed more than once and further, each entry would correspond to a functioning hospital. In short, hospitals that had gone out of business would not be included on the list. The auxiliary information would be used for stratification to improve the precision of survey estimates.

Of course, this ideal frame will never be found. To increase the accuracy of survey estimates, however, the survey planner should construct a frame as close to the ideal as possible. Whenever possible, the frame should be based upon an actual list of population elements or sampling units to which the population elements can be uniquely linked. This may not be possible when the survey focuses on households or persons since accurate lists may not be available. In this situation, an area frame is used where the sampling units are geographical areas. The methods that can be used to construct a frame as similar as possible to the ideal frame depends upon the type of frame being developed. In many cases, procedures are available that may compensate for frame deviations from the ideal.

2.4.1 List Frames

Frequently, health care surveys can be designed using a list frame. When the population of interest is health care providers, for instance, state licensing boards can often provide recent lists of licensed practitioners. The National Master Facility Inventory, available from the National Center for Health Statistics (1981, p. 26), is a list of the facilities in the United States that provide medical, nursing, personal care, or custodial care to groups of unrelated individuals on an inpatient basis. Health agency benefit records and hospital discharge summaries are other lists that may form the frame for a health care survey. In constructing a list frame, the survey planner should evaluate the available lists with respect to the impact of deviations from the ideal.

The extent to which members of the target population are missing from the frame is referred to as _undercoverage_. Depending upon the amount of undercoverage, the creditability of survey estimates can be seriously affected. The most apparent problem is that population counts will be underestimated to the same extent that population members are missing from the frame. Other population totals will be underestimated but to an unknown degree. Survey estimates such as means or proportions will also be biased in an unknown degree and direction.

As an example, lists of November 1979 Medicaid beneficiaries were used to construct frames for the State Medicaid Household Surveys (SMHS) component of NMCUES. The target population for each of the state surveys was the universe of Medicaid households within the state as of November 1979 that contained at least one civilian, noninstitutionalized Medicaid beneficiary. Since the frame files were extracted from the November 1979 payment files for the four states, they are essentially complete with respect to containing all Medicaid recipients. However, since the SMHS collected data for all of 1980, some analysts may be tempted to use SMHS to report about all 1980 Medicaid recipients in these states. The SMHS surveys will be subject to undercoverage with respect to this 1980 Medicaid population. For instance, the count of Medicaid recipients will actually estimate the total number of Medicaid recipients who also received benefits in November 1979 or who are related to November 1979 beneficiaries; the 1980 beneficiary count will be underestimated to the extent that post-November-1979 beneficiaries were not members of pre-existing Medicaid households. The total medical expenses for 1980 Medicaid beneficiaries will also be underestimated; the bias in the survey statistic will be dependent upon how the total medical expenses of noncovered beneficiaries compare to the expenses of all 1980 beneficiaries. Means and proportions will be biased with the amount depending upon the degree of undercoverage and the extent to which noncovered beneficiaries have different patterns of expenses from covered beneficiaries. Since SMHS was never intended to survey all 1980 beneficiaries, this undercoverage is to be expected.

Undercoverage bias derives from the fact that only a subset of the target population can be surveyed using the frame. The best method to reduce this bias is to develop a supplementary mechanism that identifies noncovered population members and allows them to be added to the sampled population. Frequently, interviewing individuals not included in the sample frame is more costly and hence population members not covered by the frame are included with smaller probability than frame members. For the previous example of SMHS, recent beneficiaries could have been included in the survey by periodically requesting data for all beneficiaries entering the program after November 1979 from a sample of counties and then including a subsample of these beneficiaries in the survey. This procedure was not implemented for SMHS because of the burden that would result for the states.

Usually a mechanism can be developed that will reduce the undercoverage bias, but it may be too costly in terms of time or money to implement. In this situation, the researcher has several options. Outside counts of population totals can sometimes be obtained that are not subject (or are less subject) to coverage bias. These counts can be used to adjust the sampling weights so that the undercount in population totals is reduced. This method is described in more detail in Chapter 7. The second option is to do nothing to supplement the frame nor to adjust survey estimates to reduce the bias. When the undercoverage rate is low, this may be an acceptable approach. The approach used by SMHS was to restrict the target population to a meaningful population for which a complete frame was available. Regardless of the method used, the researcher should assess the extent of undercoverage and report on its projected impact on survey estimates.

It is an unusual frame that does not miss some members of the target population; it is also common for the frame to contain listings that are not members of the target population. This latter frame deficiency is much easier to deal with, however. Listings that do not correspond to a member of the target population are referred to as ineligible or out of scope listings. The preferred strategy when the frame contains ineligible listings is to remove all ineligible listings prior to sample selection. Frequently, the list will contain auxiliary variables that can be used to remove many ineligibles.

Screening may still be needed during collection to establish eligibility. For instance, in using the state lists of Medicaid beneficiaries as the sampling frame for the SMHS, all identifiable Medicaid cases composed only of ineligible individuals in nursing homes, personal care, or custodial care institutions were removed from the lists prior to sample selection. The resultant lists still contained some ineligible cases. Cases whose members had entered institutions in December 1979 are an example of out of scope units that were contained in the frame and not identified prior to sample selection.

When proper survey procedures are used, the presence of out of scope units need not introduce bias in survey estimates. When a sampling listing is identified as out of scope, data from that listing are not used in forming survey estimates. Control system information that indicates that the listing is ineligible is recorded. No substitutions are made for the ineligible listing and the interview is terminated. Using this procedure, lists containing out of scope units can be used as sample frames without producing biased survey estimates. However, out of scope units are a form of frame inefficiency and should be removed whenever possible since they increase the survey cost per completed interview and can decrease the precision of survey estimates.

Another frequently occurring event is <u>clustering</u>, frame listings that correspond to more than one member of the target population. The SMHS furnishes an example of this event since Medicaid beneficiaries were surveyed using a list of Medicaid cases as the frame. The cases on this list were frequently associated with households containing more than one person. An Aid to Dependent Children (AFDC) case usually contained the mother and the minor children in the household, for instance. The general procedure used in this situation is to include in the survey all eligible individuals associated with each sample listing. An alternate approach is to randomly select one individual from each sample cluster for inclusion in the survey. If weights are used that reflect the fact that the probability of selection for an individual depends upon the number of individuals in his cluster, sample estimates will be unbiased. However, the resultant unequal weights may also increase the variance of survey estimates.

To the extent to which individuals within a cluster have similar responses, the effect of clustering is to increase the variance of survey estimates over that which would have been achieved with simple random sampling. Generally, clusters should be broken apart prior to sample selection so that each frame listing corresponds to only one member of the target population. However, cost savings associated with data collection within clusters may allow sample size increases that more than offset the increased variance due to clustering. In addition, some surveys have more than one unit of analysis and the cluster may be one of these units. For instance, many national surveys include household-level analyses as well as person-level analyses. In this situation, entire clusters are sampled to permit cluster-level analyses.

For effective use as a frame, a list must also contain accurate information that allows each listing to be identified and contacted for interview. When addresses are found to be incorrect, the usual procedure is to trace the individual to obtain the correct address and complete the interview. Thus, inaccurate address data increase the cost of data collection. Since the Post Office maintains change of address information for only one year, older addresses may result in sample selections that are difficult to trace. Untraceable individuals increase the total amount of nonresponse to the survey. As with undercoverage, nonresponse results in part of the target population being unrepresented by the sample. However, unlike undercoverage, the extent of the nonresponse is quantifiable using the sample data. To reduce the bias caused by survey nonresponse, nonresponse and post-stratification adjustments can be made to the sample weights. These weighting adjustments are discussed in Chapter 7.

2.4.2 Area Frames

Frequently, a list of target population members is unavailable and it is cost prohibitive to create one for the entire population. In this situation, the target population can frequently be linked to areas for which maps are available. A probability sample of these areas can then be selected. Within the sample areas, target population members are enumerated and sample members selected. The frame errors discussed in the previous section are also applicable for area frames.

When area frames are used, a multi-stage clustered sample is typically used. This clustering has the effect of increasing the variance of survey estimates over that which would have been achieved by a simple random sample of the same size. Even when a population list is available, it may still be advantageous to use area sampling. To the extent that the list is inaccurate, undercoverage bias will result which can only be partially compensated for with the measures described earlier. Since the variance can be controlled by sample size and the bias cannot, area sampling may be preferrable to using an inaccurate list.

Personal interview surveys, such as the NMCES and NMCUES household surveys, are the most common instances in which area frames are used. Areas are defined which are usually based upon geographical units used by the Bureau of the Census. To avoid frame multiplicity, each individual is linked to his usual place of residence and interviewed only if the residence is selected. Undercoverage will occur for individuals with no usual place of residence such as transients and street people.

Another source of undercoverage is associated with the failure to list all dwelling units that contain target population members. Units that are vacant when sampled areas are enumerated may be occupied at the time of data collection. Seemingly ineligible units such as stores or commercial enterprises may have an apartment attached where the owner or caretaker resides. These two sources of undercoverage are best handled by enumerating all building units within the sampled areas and selecting from the list of building units. Units that do not contain population members at the time of data collection will be ineligible for the survey. The treatment of ineligible units and the effect on sample estimates are the same as that described for list frames.

Undercoverage also results from the failure to list all building units within the sample areas. Basement apartments and apartments over garages are frequently missed. When there is a time lag between enumeration and interviewing, new construction can also result in unlisted units. With effective field procedures, many of these missed units can be identified during data collection. In this situation, the half-open interval technique can be used to give missed units a chance to be selected. The half-open interval technique links each missed unit to the unit immediately preceding it in the list of building units ordered by address. When the immediately preceding unit is selected for the sample, the missed unit is included as well. This effectively eliminates the undercoverage bias associated with missed units that are identified during field operations. This bias reduction is achieved at the expense of a small increase in the variance of sample estimates.

2.5 Sample Selection Procedures

The typical national household survey has a complex design structure with several stages of sampling for which separate frames may be required and different types of sampling procedures used. It is beyond the scope of this book to attempt to discuss these design types and sampling procedures in sufficient detail that the reader could implement them. Instead, design components that occur in most sampling plans will be reviewed in the remainder of this chapter. For additional information about the design of sample surveys, the reader should consult standard sampling textbooks such as Cochran (1977), Kish (1965), or Levy and Lemeshow (1980).

2.5.1 Simple Random Sampling

Perhaps the easiest method of sample selection is simple random sampling, which selects n units out of a population of N units in such a way that (1) no unit is selected more than once and (2) each of the distinct samples has an equal chance of being selected. Using this method, the units in the population are numbered from 1 to N. Random numbers between 1 and N determine the n selections in the following manner. The first selection is determined by the first random number. Random numbers are then used successively to determine the remaining selections so that at each draw an equal chance of selection is given to each unit of the population not already selected. At each draw, the next selection is determined by the random number as long as the number does not correspond to a previous selection. If the random number corresponds to a previous selection, the number is discarded and the next random number is examined. The process is continued until n selections have been made. Since the population from which each successive selection is made is the total population of N units minus those units already selected, simple random sampling is also referred to as *random sampling without replacement*.

When simple random sampling is used, the mean is estimated as the numeric average of the n observations or

$$\bar{y}_{srs} = \sum_{i=1}^{n} Y_i \,/\, n \tag{2-1}$$

The variance of the simple random sampling mean can be estimated as

$$\mathrm{Var}(\bar{y}_{srs}) = (1-f)\; s^2 \,/\, n \tag{2-2}$$

where $f = n/N$ is the sampling fraction and s^2 is the estimated population variance or

$$s^2 = \sum_{i=1}^{n} (Y_i - \bar{y}_{srs})^2 \,/\, (n-1) \tag{2-3}$$

The term (1-f) in equation (2-2) is referred to as the finite population correction factor. As the population size increases, the finite population correction factor approaches one and can be ignored for large populations.

Simple random sampling is not commonly used as the selection method for sample surveys for a number of reasons. Use of the method requires that a list of population elements be available, which is not the case for many populations. Even when lists are available, it may be more convenient to use a sequential selection method such as systematic sampling. The primary reason that simple random sampling is not commonly used is that it is often not cost effective. Techniques such as area sampling or stratified sampling frequently result in samples with equal precision for lower costs. However, simple random sampling is commonly used as a basis of comparison of the efficiency of other sampling techniques.

2.5.2 Stratified Sampling

Frequently, the sampling frame contains ancillary information about the population elements that can be used directly in sample selection to increase the precision of survey estimates or to provide needed control over the distribution of the sample. For instance, the list frames of Medicaid cases used in selecting the SMHS samples contained information about the type of benefits received, whether aged, disability, aid for dependent children, or state only benefits. Area frames are often composed of Census units for which socio-demographic characteristics are available.

These data may be used to partition the target population into groups, or strata. When sampling is implemented independently within these strata, the sample is said to be stratified. In constructing survey estimates, sample statistics are first computed for each stratum. Weighted combinations of these stratum statistics are then used to form the survey estimate of the population parameter and its associated variance. The weights are functions of the size of the stratum and the size of the entire population.

To illustrate the estimation procedures associated with stratified samples, suppose that a sample of size n_h is selected from the h-th stratum (h = 1,2,...,H) where the entire frame has been partitioned into H strata. With a total sample size of n units, the stratified sample estimate, $\bar{y}_{str}$, for the population mean $\bar{Y}$ can be obtained as the weighted sum of the stratum means, or

$$\bar{y}_{str} = \sum_{h=1}^{H} W_h \bar{y}_h \tag{2-4}$$

where $\bar{y}_h$ is the sample estimate of the stratum h mean, $\bar{Y}_h$, and W_h is the weight associated with stratum h. The stratum means are weighted by the fraction of the population falling into the stratum; that is, the weight W_h will equal N_h/N where N_h is the size of the h-th stratum and N is the total population size. The variance of the sample estimate can be obtained as

$$\text{Var}(\bar{y}_{str}) = \sum_{h=1}^{H} W_h^2 \text{Var}(\bar{y}_h) \tag{2-5}$$

where $\text{Var}(\bar{y}_h)$ is the variance of the stratum h mean. When simple random sampling is used to select the sample from each stratum, equations (2-2) and (2-3) are used to calculate the variance of the stratum means.

To calculate the variance of stratified mean, the within-stratum variance must be estimable which implies that at least two units must be selected from each stratum. It is not uncommon, however, for the sample to be deeply stratified with only one selection made per stratum. In this case, an estimate of the variance may be obtained by collapsing strata. The strata are grouped into pairs or pseudostrata based upon sample design characteristics and the variance is calculated using the stratified sampling formula and these pseudostrata. This method produces variance estimates that are conservative since the variance estimator will tend to overestimate the true variance of the sample statistic.

One reason for using stratified sampling is to increase the precision of survey estimates. The variance of survey estimates is reduced and hence the precision increased through stratified sampling when the strata are formed so as to maximize the between-stratum variation and minimize the within-stratum variation. Variables to use for stratification are those that result in homogeneity of the units within strata and heterogeneity between the stratum means.

The choice of population characteristics to use for stratification should reflect the following ideas:

- Usually more gain results from coarser division with several variables than from finer divisions with only one variable.

- There is no need for symmetry in forming cells. Very small cells may be combined.

- Different criteria may be used for different subgroups. It may be decided to partition males with respect to different characteristics than females.
- The classifying variables should be unrelated to each other. If two variables are highly correlated, then either will describe approximately the same amount of variation.
- The classifying variables should be accurately defined and known for all members of the target population.

The latter idea needs further elaboration. Incorrect stratifying information will result in misclassification of population units. Since misclassified units will have an opportunity for selection, the resultant sample will be a probability sample; however, the sample will be less efficient than it could have been with accurate classification of population units. Missing data for stratification variables presents a slightly different problem. Since sampling occurs within strata, each population member must be assigned to a stratum in order to have an opportunity for selection into the sample. To avoid undercoverage of the target population, a separate stratum can be created for population units with missing data or the units can be arbitrarily assigned to one or more strata. With either approach, the probability nature of the design will be preserved but again the efficiency of the design will be reduced to the extent to which the strata are not internally homogenous.

Another reason for using stratified sampling is to control the distribution of the sample. For instance, public acceptance of the results from a national survey is enhanced by maximizing the geographical dispersion of the sample. Stratifying by geographical proximity insures that the sample will be geographically dispersed.

When population elements can be classified with respect to their membership in key reporting domains, the precision of domain estimates can be improved by stratifying according to domain membership. Thus, not only is sample creditability enhanced by geographical stratification but also the accuracy of regional statistics is improved.

A final reason why stratified sampling is used is for operational convenience or necessity. For instance, if a list frame of population members is available for some but not all of the target population, it might be cost effective to stratify the population based upon presence in the list frame. Sample individuals selected from the list frame strata could then be surveyed using the less expensive data collection procedures available for list frames, such as mail or telephone interviewing. Unlisted population members could be identified and sampled, possibly at a lower rate, using more costly area frame sampling procedures. When oversampling of population subgroups is required to meet survey objectives, it may also be convenient to partition the population into comparable strata. A population subgroup is said to be *oversampled* when members of the subgroup are included in the sample at a higher rate than other subgroups. Oversampling is typically used to increase the sample size for small domains that might otherwise have sample sizes too small to analyze.

The sample designs for the NMCES and NMCUES national household surveys illustrate these reasons for using stratified sampling. Both surveys were based upon the national general purpose area samples of the Research Triangle Institute (RTI) and the National Opinion Research Center (NORC). Prior to selection of the two general purpose samples, the land area of the United States was stratified based upon Census data such as region, metropolitan status, and urban versus rural characteristics. The primary purpose of the stratification was to disperse the sample geographically since estimates were to be reported for geographical domains such as Census regions, metropolitan versus nonmetropolitan areas, etc. A gain in precision is to be expected from the stratification but it is likely to be small since the geographical factors are not highly correlated with health care use and expenditures. Since the cost associated with stratified sampling is minimal, even modest gains in precision make stratification a cost effective procedure. The two surveys also furnish an example of stratification for administrative purposes. The original RTI and NORC general purpose samples were for the continental portion of the United States, while the target population of NMCES/NMCUES included residents of Alaska and Hawaii as well. To provide coverage of all 50 states, the RTI and NORC general purpose samples were supplemented by creating a special stratum for Alaska and Hawaii and making independent selections from this stratum.

2.5.3 Area Samples and Clustering

Cluster sampling is a common feature of many sample designs as the result of cost savings that can be obtained by clustering the sample rather than by a more uniform distribution of the sample across the population. This is particularly true for area household samples where multistage sampling procedures are typically used.

The NMCES and NMCUES sample designs furnish an example of area sampling procedures that are commonly used. The first stage of sampling for both designs involved the construction of a first stage sample frame in which the entire land area of the United States was partitioned in primary sampling units (PSUs). These PSUs were defined in terms of counties or groups of counties. A first stage sample of PSUs was then selected by stratified sampling where the strata were defined based upon geographical and demographic data that were available on the Census data tapes. The sample PSUs were partitioned into second stage units (SSUs) which were defined in terms of Census-defined enumeration districts and block groups; second stage sample selections were made from each sampled PSU. At this point, lists of building units within the sample SSUs were developed. The next stage of sampling involved the selection of sample building units from these SSUs. All eligible residents of the sample building units were included in the household survey.

Two advantages accrue from this area sampling approach. The first advantage is that list frames of target population members need be developed for the sample areas only. No household survey could afford to prepare such lists for the entire nation. A second advantage is that interviewing activity is also restricted to the sample areas, reducing travel time and cost. Even when list frames of target population members are available, geographical clustering of the sample can still be cost effective. For instance, using a list frame of Medicaid cases, the SMHS samples were clustered based upon ZIP Code area to reduce geographical dispersion.

The disadvantage of cluster sampling approaches is that the variance of the cluster sample is increased over the variance that would have been achieved had simple random sampling been used with the same sample size. The increased variance is the result of the tendency of physically clustered units to have homogeneous responses. This within cluster homogeneity is measured by the _intra-cluster correlation coefficient_, which is the correlation coefficient ρ between units from the same cluster.

Ignoring the covariances arising from sampling without replacement from a finite universe, the sampling variance of a mean, $\bar{Y}$, estimated from a cluster sample can be approximated as

$$\mathrm{Var}(\bar{y}_{c\ell s}) = \frac{\sigma^2}{n}\,[1+(b-1)\rho] \tag{2-6}$$

where σ^2 is the population variance, n is the total sample size, and b is the average cluster size. The range of values that ρ can assume is $-1/(b-1) \leq \rho \leq 1$. Negative values for ρ will occur when the units within each cluster tend to be representative of the population of units. When $\rho = 0$, the variance is the same as that which would result from simple random sampling. For area cluster samples, ρ is generally a small positive number. Even with a small ρ value, the variance of a clustered sample can be large relative to that of a simple random sample when b is large. If $\rho = 0.1$ and $b = 21$, for instance, the cluster sample variance will be 20 percent greater than the simple random sampling variance.

The cost effectiveness of a clustered design can be assessed by comparing the cost to that of a simple random sample with the same effective sample size. The _effective sample size_ of a proposed design is the sample size required to achieve estimates of the same precision using simple random sampling from an infinite population. For clustered designs, the effective sample size is $n/[1+(b-1)\rho]$. The cost advantage of clustered designs derives from the fact that the clustered sample of size n can often be obtained at less cost than a simple random sample of size $n/[1+(b-1)\rho]$.

A final consideration for clustered designs is the optimal number (b) of sample individuals to interview per cluster, the total number ($n = ab$) of interviews to be conducted, and the number of required clusters (a). To model this problem, consider a two-stage clustered design in which a primary clusters are chosen from a total of A clusters. For each of the a clusters that are selected, b dental hygienists are randomly chosen and a personal interview conducted with each. The optimal value for these variables will depend upon the variance of the resulting estimators as well as cost factors. A model for the overall variable cost (C) when a clustered sampling plan is used is

$$C = nC_0 + aC_1 \tag{2-7}$$

where C_0 is the unit cost of an interview and C_1 is the cost per cluster. The unit cost C_0 will involve interviewer time, travel time, and data processing costs. The per cluster component C_1 will primarily involve interviewer recruitment and training costs. Using this cost model, the optimal number of interviews per cluster will be

$$Opt(b) = [C_1(1-\rho)/C_0\ \rho]^{\frac{1}{2}} \qquad (2\text{-}8)$$

General procedures for determination of the total sample size n are discussed in Section 2.6.

2.5.4 Systematic Sampling

Perhaps the most common method for sample selection is *systematic sampling*, particularly when list frames are available or a natural ordering can be developed for the population. With a sample of size n desired from a population of size N = nk, an integer between 1 and k is selected to be the random starting point r. The systematic sample consists of the r-th element in the list and every k-th element thereafter.

Kish (1965, pp. 115-117) describes a variety of systematic methods when the sample size n is not an integer divisor of the population size N. Perhaps the easiest to implement allows the sampling interval, k, to be a noninteger number. The random starting point, r, is a real number between 1 and k. The sampling fraction k is repeatedly added to r to produce n real numbers. The sample associated with these n real numbers is obtained by truncating the digits to the right of the decimal point.

In essence, systematic sampling implicitly stratifies the frame into n zones, making one selection from each zone. The distinction between systematic and stratified sampling is that with systematic sampling the selection of sample individuals is not independently implemented within strata. Instead of the k^n possible samples that would be obtained with stratified sampling of one unit from n strata of size k, only k possible samples can be obtained using systematic sampling.

Systematic sampling can also be viewed as selecting one cluster of n units from the k possible clusters contained in a population of size N = nk. Using the approach described in the previous section, the variance of survey estimates can be modeled as

$$\text{Var}(\bar{y}_{sys}) = \frac{\sigma^2}{n} [1 + (n-1)\rho] \tag{2-9}$$

Since only one cluster is selected (a=1), the cluster size and the total sample size are identical (n = b).

This analogy to cluster sampling can sometimes be exploited to improve the precision of survey estimates. To do so, the frame is hierarchally ordered so that the units within each cluster tend to be representative of the population of units. When the between cluster component to the population variance is zero, the intra-cluster correlation coefficient will be equal to its minimum value or

$$\min(\rho) = -1/(n-1) \tag{2-10}$$

For most applications, the intracluster correlation coefficient will be somewhat larger than the minimum value for several reasons. First, sources of variation may be inadvertently missed in sorting the data records. Second, exact correspondence between the frame records and particular sources of variation may not be possible.

Third, since cell sizes in the ordered frame file will vary, the systematic construction of the clusters will tend to include in the same cluster more records from large cells than from small cells. Nonetheless, considerations of operational convenience and statistical efficiency frequently suggest a systematic sampling approach.

Care should be taken in implementing such an approach since systematic sampling can also inflate the variance of survey estimates. When the units are arrayed in an order demonstrating a high degree of periodicity, ρ can be large and positive; in which case, the variance of the systematic sample mean will be much larger than that which would have been obtained had simple random sampling been used to select the n units. Such a situation will occur if one is sampling from a sine wave and k is equal to the period or a multiple of the period of the curve. With knowledge of the underlying periodicity in the list, systematic sampling can decrease the variance over that of simple random sampling. For instance, if one is again sampling from a sine wave and k is an odd multiple of 1/2 the period of the curve, then each systematic sample will estimate the population mean exactly and hence,

$$V(\bar{y}_{sys}) = 0 \tag{2-11}$$

and

$$\rho = \min(\rho) = -1/(n-1) \tag{2-12}$$

Thus, the frequency with which records are sampled in a list exhibiting periodicity can have a serious impact on the precision of survey estimates.

Without replication, the variance of estimates derived from systematic sampling is not estimable. Since the systematic sample of n observations is in essence one cluster from a population that contains k clusters, the sample provides no information about the between cluster variability. A common practice in this instance is to treat the systematic sample as if it had been derived by simple random sampling. Depending upon the sign of ρ, this method will overestimate or underestimate the true variance except for the rare instances in which $\rho = 0$.

In many situations, a successive paired difference approach will provide a better approximation of the variance of the mean. Using this approach, the observations are conceptually regarded as being derived from a stratified random sample with two units per stratum. If Y(i) represents the i-th observation from a systematic sample of n units, the successive paired difference approach would approximate the variance as

$$\mathrm{Var}[\bar{y}_{sys}] = \sum_{i=1}^{n/2} [Y(2i-1) - Y(2i)]^2 / n^2 \qquad (2\text{-}13)$$

Cochran (1977, pp. 223-226) presents this method and others to estimate the variance of estimates derived from a systematic sample. The method that is most applicable for a particular instance may not be easily discernable.

These approximate variance estimation procedures are not needed when more than one systematic sample is selected. For this reason, replicated systematic sampling has become the standard practice. The problem remains of how to distribute the sample between the number of clusters, a, and the size of each, b, to achieve the final sample size n = ab. The variance expression in this instance is

$$\mathrm{Var}(\bar{y}_{sys}) = \sigma^2/n\ [1 + (b-1)\rho]. \tag{2-14}$$

Hence, it is advantageous from a variance-reduction standpoint to use large b-values when ρ is negative. Since the cost of data collection usually depends upon the total number of selected records and not on the number of clusters, there is little point in having a large number of clusters beyond that required to produce reasonably stable variance estimates. As few as five clusters might be satisfactory, although the final determination will need to consider the periodicities in the frame file ordering.

2.5.5 An Alternate Sequential Selection Method

In using systematic sampling, the researcher may be selecting a sample that is more efficient than a simple random sample; this is not guaranteed since unknown frame file periodicities can adversely affect the efficiency of the sampling procedure. A sample selection procedure that avoids this problem while retaining the advantages of implicit stratification of an ordered list is described by Chromy (1979). When n sample units are desired, Chromy's procedure essentially divides the file into n zones and randomly selects one unit from each zone. At the first stage of sampling Medicaid households for the SMHS, for instance, Chromy's procedure was used to select the sample of area clusters defined by ZIP Codes. The size measure used in the selection was the number of Medicaid cases within the

ZIP Code area. In using Chromy's procedure to select n units with equal probability, the size measure will be uniformally equal to one.

Chromy's selection algorithm sequentially considers each unit on the ordered list and probabilistically determines the number of times each unit will be included in the sample. The sample is selected with probability proportional to size since the expected number of times a unit is selected is proportional to its size measure. The selection is also probability minimum replacement; the actual number of times a unit is selected can only assume two values - the integer portion of the expected number of selections or the next largest integer. Williams and Chromy (1980) present a computerized version of Chromy's procedure and describe its use.

In order for the variances of survey statistics to be estimable, all possible pairs of units must have a positive probability of appearing in the sample. Chromy's procedure accomplishes this condition by linking the end of list to the beginning to form a circular array and selecting a unit at random to begin the sequential selection process. By initiating the zone formation at a randomly selected point on the circular listing, adjacent units have a chance of being selected despite the single selection per zone characteristic of the selection process. Thus, Chromy's sequental selection procedure has the implicit stratification benefits of systematic selections from ordered listings while providing for unbiased variance estimation. Chromy (1981) presents variance estimators for the design.

2.6 Determination of the Total Sample Size

The remaining design consideration to be discussed is the determination of the sample size for a survey. The sample size will be a function of the analyses that are planned and the required precision of survey estimates. Most survey data analyses involve the computation of means, proportions, and cross tabulations and the comparison of estimates for population subgroups. By modeling the precision of these survey estimates, the required sample size can be determined.

Examining the precision of survey estimates for the range of possible values of the parameter being estimated can provide an assessment of the sample size requirement. An approach that is frequently helpful is to calculate the *coefficient of variation* of the estimate, defined as the standard error of the estimate divided by the parameter being estimated. The survey objectives should be examined to determine the value that is required for the coefficient of variation of survey estimates. Generally, the desired value for the coefficient of variation will range from 0.05 to 0.10, since larger values may result in confidence intervals about estimates that are so broad as to be of limited utility. In order to use the coefficient of variation criterion, information is needed about the magnitude of the standard error. This information may be obtained from studies reported in the literature or by modeling the estimate and its standard error.

Many questionnaires are composed of items with categorical responses primarily, with continuous responses obtained for only a few items. In this situation, sample size requirements can be modeled by examining the precision of survey-estimated proportions. The variance of an estimated proportion can be expressed as a function of the parameter being estimated. Ignoring finite population correction factors, the sampling variance of an estimated proportion, $\hat{P}$, may be expressed as

$$\text{Var}(\hat{P}) = [P(1-P)/n]\ \text{DEFF} \qquad (2\text{-}15)$$

where n is the sample size, P is the proportion being estimated, and DEFF is the design effect. The _design effect_ for a survey estimate is defined to be the ratio of the variance of the statistic under the actual design divided by the variance that would have been obtained from a simple random sample of the same size. The design effect represents the cumulative effect of such design components as stratification, unequal weighting, and clustering and will differ for each design. Past survey experience is usually the source of design effect estimates as statistical literature provides little guidance.

Another way to determine the required sample size is to examine minimum detectable differences for important domain comparisons. Suppose that the survey objectives require that hypotheses such as the following be tested:

$$H_o: \ P_1 = P_2$$

$$H_A: \ P_1 > P_2$$

Here P_1 represents the proportion of members of domain 1 who respond in a certain manner. P_2 has a similar definition for domain 2. The values of P_1 and P_2 are estimated from the samples drawn from the two domains. The estimator $\hat{P}_i$ for the i-th domain (i=1,2) is distributed binomially with mean P_i and variance $d_i P_i (1-P_i)/n_i$, where n_i is the domain sample size and d_i is the design effect.

In testing the hypothesis, it is convenient to transform the binomial proportions to

$$\hat{T}_i = 2 \arcsin (\hat{P}_i^{\frac{1}{2}}) \tag{2-16}$$

since the transformed variable is approximately normally distributed when the sample size is sufficiently large (Snedecor and Cochran, 1980, pp. 290-291). The transformed proportion has as its mean

$$T_i = 2 \arcsin (P_i^{\frac{1}{2}}) \tag{2-17}$$

and variance

$$\text{Var}(\hat{T}_i) = d_i/n_i \tag{2-18}$$

If the null hypothesis is true, the statistic $\hat{T}_1 - \hat{T}_2$ is distributed about zero; otherwise the statistic is distributed about $T_1 - T_2$ which is greater than zero.

The minimum difference (Δ) that can be detected with probability $1 - \beta$ when testing with significance level α can be shown to be

$$\Delta = (Z_\alpha + Z_\beta) (d_1/n_1 + d_2/n_2)^{\frac{1}{2}} \tag{2-19}$$

with Z_α and Z_β the critical points of a standard normal distribution where the values of the cumulative distribution function are $1-\alpha$ and $1-\beta$ respectively. To convert Δ which is in terms of the minimum detectable difference between the transformed proportions to the minimum detectable difference between the untransformed proportions, either the value of P_1 or of P_2 must be specified since a difference of say Δ' between P_1 and P_2 is not equivalent to a fixed value of Δ. To see this, note that a minimum detectable difference of $\Delta = 0.25$ translates to a difference of 0.02 between the proportions P_1 and P_2 when $P_2 = 0$ and to a difference of 0.36 when $P_2 = 0.50$.

These and other approaches can be implemented to determine the required sample size. The most important aspect of the sample size determination, however, is carefully examining the survey objectives to determine the types of anaysis that will be required, the reporting domains for the survey, and the required precision of survey estimates.

References

Chromy, J. R. (1979). Sequential Sample Selection Methods. Proceedings of the American Statistical Association, Survey Research Methods Section, 401-406.

Chromy, J. R. (1981). Variance Estimators for a Sequential Sample Selection Procedure. Current Topics in Survey Sampling, editors: D. Krewski, J.N.K. Rao, and R. Platek, New York: Academic Press, 329-347.

Cochran, W. G. (1977). Sampling Techniques, New York: John Wiley and Sons.

Cohen, S. B. and W. D. Kalsbeek (1981). NMCES Estimation and Sampling Variances in the Household Survey, NHCES Instruments and Procedures 2, DHHS Publication No. (PHS) 81-3281, Washington, DC: U.S. Government Printing Office.

Cox, B. G., L. L. Piper, and R. E. Folsom (1983). NMCUES Survey Design and Methodology, NCHS Series II Report (in press).

Deming, W. E. (1960). Sample Design in Business Research, New York: John Wiley and Sons.

Kish, L. (1965). Survey Sampling, New York: John Wiley and Sons.

Kotz, S. and N. L. Johnson (1982). Encyclopedia of Statistical Sciences, Volume I, New York: John Wiley and Sons.

Levy, P. S. and S. Lemeshow (1980). Sampling for Health Professionals, Belmont, California: Lifetime Learning Publications.

Muse, D. N. and D. Sawyer (1982). The Medicare and Medicaid Data Book, 1981, Department of Health and Human Services, Health Care Financing Administration, Office of Research and Demonstrations, Baltimore, Maryland.

National Center for Health Statistics, N. D. Pierce (1981). Data Systems of the National Center for Health Statistics. Vital and Health Statistics, Series 1, No. 16, DHHS Publication No. (PHS) 82-1318, Washington, DC: U.S. Government Printing Office.

Rosander, A. C. (1977). Case Studies in Sample Design, New York: Marcel Dekker, Inc.

Snedecor, G. W. and W. G. Cochran (1980). Statistical Methods, 7th ed., Ames, Iowa: The Iowa State University Press.

Williams, R. L. and J. R. Chromy (1980). SAS Sample Selection Macros. Proceedings of the Fifth Annual SAS Users Group International Conference, 392-396.

3
A Comparison of Two Longitudinal Design Alternatives

The typical survey interviews sample individuals on only one occasion to obtain survey data. These surveys are referred to as cross-sectional since the survey data reflect the characteristics of the target population at a particular moment in time. When more than one interview is used to obtain information from a sample individual or the occupants of a sample dwelling unit, the sample design is said to be longitudinal.

Repeated interviewing may be advantageous when survey objectives involve the study of population changes. An example is the Current Population Survey (CPS), a longitudinal study that monitors change over time in the labor force status of the United States population (Bureau of the Census, 1978). Repeated interviewing allows the CPS to study dynamics of the labor market such as the length of unemployment. In addition, a composite estimation procedure can be used that improves the precision of estimates of means and change over time.

Studies that require retrospective reporting for a long period of time frequently require repeated interviews to minimize recall error. For instance, obtaining accurate annual health care data is difficult when only one retrospective interview is used (National Center for Health Statistics, 1965a, 1965b, 1972). The National Medical Care Expenditure Survey used six rounds of data collection to obtain annual health care utilization and expenditure data. The number of required data collection rounds for NMCES was determined based upon the results of methodological studies and cost and time constraints imposed on data collection and processing.

Two varieties of longitudinal designs are in common use; these varieties may be referred to as longitudinal dwelling unit designs and longitudinal household designs. Over repeated rounds of data collection, the *longitudinal dwelling unit design* interviews the current occupants of a sample of dwelling units. Thus, the basic sampling unit for the survey is the dwelling unit rather than the individual or household. In contrast, the *longitudinal household design* has the individual as its basic sampling unit, rather than the dwelling unit. Sample individuals are identified in the first data collection round as the permanent residents of the sample dwelling units. In all data collection rounds after the first, the households containing these sample individuals are interviewed, regardless of whether or not they resided in the original sampled dwelling units. The type of longitudinal design that is preferred for a particular study depends upon the analytic goals of the study and cost/precision tradeoffs associated with the designs.

Often the analytical goals of a study dictate the consideration of a longitudinal household design. This design is essential for providing accurate measurements of variations in population characteristics that occur over time. In the NMCES, many analyses concentrated on the extent of and changes in health insurance coverage over time, and the cost of illness for various diagnoses which are sensitive to seasonal change (Chapter 1). In addition, annual distributional estimates of the health care utilization and related expenditures were required. The reduction in effective sample size for these analyses that would be incurred by a dwelling unit design, in addition to potential bias in estimates due to the missing data of movers, might be unacceptable for these analytical needs.

The longitudinal dwelling unit design is the most straightforward to implement since at each round of data collection the interviewer returns to the same addresses that were visited in the previous rounds; depending upon the application, returning to the same unit may result in lower costs for data collection. With respect to the estimation of population means and totals, the longitudinal dwelling unit design is subject to less undercoverage than the longitudinal household design. In some instances, the dwelling unit may also be the unit of analysis as well. Studies that focus on characteristics of the dwelling unit or its occupants and how they vary with time may best be served by a longitudinal dwelling unit design. Examples of longitudinal dwelling unit designs include the Current Population Survey and the National Crime Survey.

Longitudinal household designs are more difficult to implement since households and persons may move during the course of the survey. Movers have to be traced and interviewed at their new location with this design. However, longitudinal data for persons or households may be needed in order to define the statistics that measure the parameter under investigation in a study. For instance, aggregates of annual medical care utilization and expenditures for sample individuals are used to investigate the characteristics of the high usage population. Studies of the high use population are important since a small fraction of the total population accounts for a large proportion of total medical care use. The upper five percentile of the national population in terms of annual medical care charges accounts for 56 percent of all medical care charges. Some analyses of changes over time also require longitudinal data. An analysis of Medicaid program participation might want to measure the rate at which households enter or leave the program. Finally, distributional studies of the population may require that data be aggregated across time in order to form the categories of the distribution. An example of this type of analysis would be categorizing the national population in terms of the number of days during the time period in which sickness forced them to restrict their normal activities.

The National Medical Care Expenditure Survey (NMCES) was a longitudinal household design measuring the 1977 health care expenditures and utilization of the civilian, noninstitutional population of the United States. In planning the 1980 sequel to the NMCES, a longitudinal dwelling unit design was investigated as an alternative design for the study. It seemed reasonable to suppose that defining the basic sampling units to be the dwelling units themselves, rather than the households residing within the dwelling units at Round 1, would decrease data collection costs for the future national medical care survey since movers would not have to be traced and interviewed at their new location. Since some planned analyses required longitudinal person or household data, the cost savings needed to be substantial to consider implementing a longitudinal dwelling unit design. For this reason, a study was conducted to explore in detail the implications of defining the sampling units to be the sample dwelling units rather than the households (families) within these housing units (Cox, et al, 1979). The activities involved in this investigation may be summarized as

- developing precise definitions for longitudinal dwelling unit and household designs,
- discussing the impact on analysis goals of dwelling unit versus household as the sampling unit,
- comparing the cost for survey operations and data processing of a dwelling unit design to that of the household design,
- computing mobility rates for subpopulations, and
- comparing annual utilization/expenditure distributions for nonmovers versus the entire population.

To capitalize on the amount of information available from the NMCES, the two designs were compared by contrasting the actual NMCES experience with results that were projected for the survey had a longitudinal dwelling unit design been used.

3.1 Comparison Between the Definition of the NMCES Household Design and a Dwelling Unit Design

The NMCES had as its target population the civilian, noninstitutionalized residents of the United States in 1977. Data were collected for only those time periods in 1977 when the sample individuals were civilian, noninstitutionalized, and residents. This constitutes a dynamic definition of the target population in the sense that membership in the target population varies over time. To reflect the dynamic nature of the target population, NMCES included as sample individuals all individuals residing within the sampled dwelling units in Round 1 and all individuals who joined these families by birth or by return from the military, an institution, or out of the country. To prevent double counting and facilitate family-level analyses, college students living away from home were linked to their home rather than college address for sample selection purposes. These sample individuals were referred to as "key" in the sense that data were to be gathered for the entire 1977 period in which they were eligible for the survey. Since family-level analyses were planned, individuals who moved into families containing these key individuals were also included in the survey but only for the time period in which they belonged to families containing key individuals. In attempting to reflect the dynamic 1977 population, NMCES did not have a mechanism whereby individuals entering the target population after Round 1 but not joining a preexisting household could be included in the survey. This design flaw constituted a source of undercoverage for individual-level and family-level analyses; the effect of the undercoverage on survey estimates is not expected to be large.

In the household interview portion of NMCES, respondents were interviewed six times at approximately twelve-week intervals beginning in early 1977 and ending in mid-1978. Information was collected on demographic characteristics and 1977 health care utilization and expenditures in the first five rounds of interviewing with a sixth round in mid-1978 to correct incomplete Medical Provider Survey permission forms and obtain other survey data. The first, second, and fifth household interviews were usually conducted in person; the other three were administered by telephone whenever possible. With the exception of data missing due to sample attrition, the design obtains complete annual data for individual-level and family-level analyses.

In contrast to the longitudinal household design, the dwelling unit design would select a sample of dwelling units and define the sample for each round to be those civilian, noninstitutionalized individuals currently occupying the selected dwelling units during that round of data collection. Thus, the dwelling unit design would treat the rounds as separate surveys and define the sample population for each round to be the civilian, noninstitutionalized population at that point in time. To preserve the clustered nature of the dwelling unit design, college students living away from home would be interviewed when their college address was selected rather than their home address. The dwelling unit approach would result in better coverage for individuals who become eligible during the year but were ineligible in Round 1. For analyses requiring annual data, the design is less satisfactory, however, since annual data will be unavailable for the approximately 20 percent of the population that move.

Some of the ways in which this alternative housing unit design differs from the NMCES longitudinal household design can be best illustrated by detailing the activities that would differ for the dwelling unit design. During Round 1, survey activity would be basically the same for both designs except for the treatment of college students living away from home. In subsequent rounds of the survey, the dwelling unit design would differ from the NMCES design in that

- individuals or families who moved out of the sample dwelling units or group quarters would not be traced and followed for interview, and
- individuals or families who moved into the sample dwelling units or group quarters would be included in the survey.

Note that when the longitudinal dwelling unit design is used, each sample listing must be returned to in each round to interview all eligible respondents residing in the unit. This means that units that were vacant in Round 1 or not being used as dwelling units would have to be checked at every subsequent round to determine their occupancy status. New construction would also be surveyed in each round using the half-open interval method, or some similar procedure. Dwelling units whose occupants refused to participate in a previous round would be checked as well as those units in which no eligible respondents were at home or available for interview in previous rounds. This is in contrast to the longitudinal household design in which the interviewers visit all sample listings in Round 1 but in subsequent rounds visit only the approximately 80 percent that were occupied by eligible individuals who agreed to participate in the survey.

3.2 Implications for Analysis of Dwelling Unit Versus Household as the Sampling Unit

The decision to choose dwelling unit or household as the sampling unit for a survey must be made in the context of the types of analyses being planned for the design. One can summarize the dilemma the survey planner has by stating that if the estimation of univariate means, proportions, and totals are of major importance, then the planner should select the dwelling unit design. However, if the estimation of distributions is more important or other analyses requiring annual data, then the household should be used as the sampling unit. Of course, in almost all surveys of a longitudinal nature, both types of data analyses are of interest. In which case, the survey designer must make a decision based not only upon a cost comparison of the two designs but also upon the types of analyses being planned, their relative importance, and the efficiency of the designs with respect to these analyses. To clarify this comment, the analytical implications of the NMCES household design will be contrasted with that of a dwelling unit design.

To compare the two designs, a distinction needs to be made between the definitions of the target population and the sample population for a survey. The _target population_ is defined to be the population about which inferences are to be made. The _sample population_ for the survey is the population from which the sample was selected. To understand this distinction, suppose results from the NMCES survey are used to estimate the total expenditures for medical care incurred by the civilian, noninstitutionalized population of the United States during the 1977 calendar year; that is, these individuals form the target population for the desired estimate of total expenditures. The sample population for NMCES was composed of those residents who were civilian and noninstitutionalized during the first round of data collection in 1977 or who were born to residents that were civilian and noninstitutionalized during the first round or who returned from institutions, the military, or out of the country and joined sample households. If NMCES data were used to obtain the above mentioned estimates, then the sample population upon which the estimates were based would be a subset of the population about which inferences were being made. Individuals who were either noncivilian, institutionalized, or nonresident during Round 1 but who became eligible in subsequent rounds and did not join preexisting households did not have an opportunity to be included in the survey. Estimates of total expenditures for this particular target population may be biased when the NMCES sample is used unless suitable adjustments are made in the estimates. In contrast to the NMCES household design, the dwelling unit design examines eligibility in every round of the survey so that the sample population approximates the target population defined above. Hence, the dwelling unit design may produce estimates with less bias than those produced by the household design when the target population is the collection of all persons who could be classified as civilian, noninstitutionalized residents at any time during the data collection period.

An entirely different situation is posed when distributional statistics concerning annual levels of expenditures and utilization are being estimated. In this situation, the target population might be all individuals who were civilian, noninstitutionalized residents throughout 1977. In this case, the target population would be a subset of the NMCES sample population. In estimating the distribution of the target population with respect to various levels of annual expenditures or utilization, the NMCES household design is ideal since civilian, noninstitutionalized residents are followed throughout the data collection period to obtain complete records of annual data. Aside from the bias caused by initial nonresponse and subsequent attrition, which can be partially removed using nonresponse adjustment procedures, approximately unbiased estimates for annual utilization and expenditure statistics can be formed directly from the NMCES data.

Because the dwelling unit design surveys the occupants of sample dwelling units during successive data collection rounds and does not follow movers, complete annual data would be available for only 70 to 75 percent of the sample individuals. This is in contrast to NMCES in which complete annual data are available for 89 percent of the sample individuals who participated in the Round 1 interview. Many analyses will require annual data in order to characterize individuals by their medical care experience (e.g., proportion of persons with large medical care expenses or persons ever participating in the Medicaid program during the reference period). To form these estimates when the dwelling unit design is used, the analyst must either impute for missing data or else form estimates based upon nonmovers and then use these results to make inferences to the total population. In addition to the bias of annual statistics, the reduced sample size will generally result in an increase in the variance of sample estimates so that the mean square error of estimates of distribution patterns for annual expenditures and utilization may be significantly greater for the dwelling unit design than the household design. The dwelling unit design will result in more than one-fourth of the sample having missing data and hence the results of any analysis will be open to criticism and not easily defensible.

If the dwelling unit design is used and distributional estimates are of importance, then it probably would be necessary to impute for incomplete data. In applying imputation procedures, one needs to recognize that such data manufacturing necessarily introduces inconsistencies into the records and may lead to less accurate estimates than would result from no imputation. The extent to which an imputation procedure can reduce the bias due to missing data depends upon how effective the demographic and other classifier variables are at predicting response. Although imputation may remove some of the bias due to missing data, there is no way of knowing how effective the imputation procedure actually is in removing the bias. Thus, if the estimation of distributions of annual medical care expenditures and utilization is a critical ingredient of the analysis plan, a dwelling unit design may not be appropriate.

Based upon the types of analyses being planned for the data resulting from future medical care surveys, a household design such as that used by NMCES was judged to be most appropriate for data collection for future surveys. The unit of analysis for most health care investigations incorporating survey data are the annual level of variables such as utilization, expenditures, or workdays lost due to sickness. For instance, the distribution of health care utilization and expenditures is of importance in determining whether certain groups, such as the poor and elderly, are receiving adequate medical treatment on a regular basis. The determination of the cost of various proposed medical care programs is another important component of the analysis plan for the health care surveys that uses annual data. In these cost determinations, the annual level of expenditures as well as the unmet medical needs of sample individuals are analyzed, taking into account the change in supply and demand expected to result from the proposed medical care programs. Since means, proportions, and totals, which are best estimated using the dwelling unit design, can be obtained without an undue level of bias using the household design, the preferred sample design for future medical care surveys would be a longitudinal design with individuals as the basic sampling unit. Unless there were substantial cost savings (for the same level of precision) associated with the use of the longitudinal dwelling unit design, there would be little reason for its use in future medical care expenditure surveys.

3.3 Cost Comparison of the Dwelling Unit Design Versus the NMCES Household Design

For this reason, an important step in assessing whether the sampling units used for future medical care expenditure surveys should be dwelling units rather than individuals was to estimate the survey operations and data processing costs of a dwelling unit design. The cost comparison of the household design used for NMCES versus the dwelling unit design included all costs for survey operations and data processing. Sample selection and data analysis costs were not expected to be different for the new design. The costs described in the remainder of this section are 1977 costs.

3.3.1 Survey Operations Cost Comparison

The first step in comparing data collection costs for the two designs was to obtain estimates of the number moving each round and how far they moved. Note that movers may be entire families that move or they may be individuals who leave their family's residence. The latter are the results of events such as marriage, divorce, and children leaving home to work or attend college elsewhere. Sample individuals were assigned to interviewers by "reporting units" where a reporting unit was composed of family members living in the same dwelling unit. When an individual left home, the composition of his old reporting unit was changed to reflect the fact that he had left and a new reporting unit was created to follow the individual and interview him at his new residence. For both individual and family moves, the interviewer was instructed to complete the interview either in person (or by telephone if necessary) when the reporting unit was located within a two-hour one-way drive. Reporting units whose new address was within a two-hour one-way drive of another primary sampling unit (PSU) were reassigned to the interviewer in that PSU for completion of the interview. Reporting units located outside a two-hour, one-way drive of any PSU were completed over the telephone by the staff assigned to a nearby PSU.

A measure of movement for sample individuals was obtained by examining NMCES change of address cards. During the initial interview, each sample household was given a change of address postcard and instructed to return the card if they moved. Postcards were returned from approximately 14 percent of the sample households during the course of the survey. The distribution of movers using the postcard data suggests that the majority of moves took place within a narrow geographical range. The movers were categorized into the following types based upon distance moved:

- move within the same city,
- move to another city in the same state, and
- move to another state.

The number of movers was then tabulated by the date given on the change of address card for the three types of moves. The results of these tabulations are presented in Table 3-1. Note that 48 percent of all moves involved a change of address within the same city. These movers would most likely be interviewed by their original interviewer and little additional interviewer travel would be required. Thirty-four percent of the moves were a change of city within the same state while only 18 percent involved a change of state. Those moves involving a change of state entail the most expense since there usually was administrative work needed in reassigning the case as well as more interviewer travel in the personal interview rounds.

Table 3-1. Summary of Household Movement Data Abstracted From NMCES Change of Address Cards

Date	Total Cards Received	Distribution by Location (%)		
		Within Same City	Different City Same State	Different State
1977:				
January - March	98	52	29	19
April - June	154	41	37	22
July - September	338	39	38	23
October - December	346	51	30	19
1978:				
January - March	266	52	34	15
April - June	220	46	35	19
July - on	73	58	30	12
Date Unknown	282	51	35	15
Total	1,777	48	34	18

Any estimate of the percent of movers in the sample computed using change of address cards would be an underestimate since not all movers submitted change of address cards and the change of address cards were only returned when entire families moved. Another estimate of the percent of movers was obtained while computing mobility rates for individuals (see Section 3.3). For this analysis, a move was defined to have occurred when the ZIP Code changed in a round from what it had been in Round 1. Using this definition of movement, 8.7 percent of the sample individuals had moved by Round 2, 14.9 percent had moved by the Round 3, 18.2 percent had moved by Round 4 and 22.3 percent had moved by Round 5. Note that these mobility rates were tabulated for individuals and not households while survey costs are largely a function of the number of household reporting units. Further, these mobility rate data do not furnish an estimate of the number of college students living away from home who required a separate interview.

For the purpose of determining how much it will cost to follow movers and family members living away from home, the distribution of movers given in Table 3-1 from the tabulations of change of address cards were used to estimate the distance individuals moved. Noting that when households move, they must be followed in that round and all subsequent rounds and using the results of the mobility calculations discussed subsequently, an estimate was made that approximately 15 percent of the NMCES interviewing occurred outside the original sampled areas. This estimate of the number of interviews with movers and/or family members living away from home and the associated geographical distribution were used in all subsequent tabulations of the cost of interviewing movers versus nonmovers.

The next step in comparing the feasibility of the dwelling unit design to the household design used for NMCES was to compute the costs actually incurred in the data collection phase of the survey. To assess the effect on field operations of following families and individuals who moved or were temporarily living away from the sampled dwelling units, average variable per unit costs for field operations were estimated. The costs were developed by partitioning NMCES costs into fixed and variable costs. _Fixed costs_ are those costs that are not dependent upon the sample size used (e.g., administration, instrument development). Variable costs are those costs that are directly related to the sample size used (e.g., interview costs, data entry).

These variable costs are presented in Table 3-2. In this table, the "Professional Labor" category refers to the cost for professional labor. The "Other Costs" category includes costs for items such as materials, shipping, telephone charges, travel, report printing, respondent incentives and interviewer services, and other expenses. "Total Technical Costs" is computed as the sum of professional labor and other direct costs. Finally, the "Administrative Expense" is a surcharge on total technical costs designed to recover overall administrative costs for personnel management. Project activities for which separate costs were prepared include: (1) recruitment of interviewers, (2) counting and listing dwelling units within sample segments, (3) interviewer training charges, (4) field operations associated with data collection activities, (5) data receipt operations in house, (6) data coding, (7) data entry or keying, and (8) data processing.

Survey operations personnel studied these unit costs and estimated the variable cost component associated with an interview with a "mover", i.e., a family or individual living outside the clustered dwelling units. The additional costs associated with interviewing a mover include tracing costs to obtain the new address, travel costs both in terms of mileage and interviewer time, and administrative costs. Tracing was typically not a time consuming task since change

Table 3-2. Estimates of the NMCES Variable Cost Experience Per Completed Interview (in 1977 dollars)

Project Activity	Cost Category				
	Professional Labor	Other Costs	Total Technical Costs	Administrative Expense	Total Costs
Recruiting	0.56	0.30	0.86	0.08	0.94
Counting/Listing	0.90	0.68	1.58	0.15	1.73
Training Charges	4.53	11.20	15.73	1.51	17.24
Field Operations	13.17	35.24	48.41	4.54	52.95
Data Receipt	0.98	7.29	8.27	0.77	9.04
Data Recoding	1.15	1.64	2.79	0.27	3.06
Data Entry	1.07	10.07	11.14	1.03	12.17
Data Processing	13.20	16.33	29.53	2.73	32.26

of address cards were sent in for approximately 40 to 50 percent of the families that moved and families usually provided address data for individuals moving out of the home. Further, the name, address, and telephone number of "someone who will always be able to locate you" were obtained in the Round 1 interview and used for tracing when no change of address data were available from the post office. The unit costs for interviews with movers are given in Table 3-3.

Using estimates of the total number of movers and these estimated variable costs for interviewing movers, the costs of interviewing movers was subtracted from the total field operations costs and the unit costs for nonmovers estimated (see Table 3-3).

The costs for interviewing nonmovers were then used as the basis for estimating the variable costs for the longitudinal dwelling unit design. Minor adjustments were made to the unit costs for non-movers in creating the longitudinal dwelling unit design costs for the following reasons:

- Off-site professional labor (field supervisors) was expected to increase per interview since more refusals would have to be handled after Round 1 (new families move into the dwelling units every round) and since personal validations would be needed of sample listings classified as vacant or not a dwelling unit (in NMCES, most validation of interviewer work could be done by telephone after Round 1). Additional field supervisor time would also be needed to advise interviewers on special problems associated with the dwelling unit design such as how to handle new construction.

- Additional materials would be needed to handle vacants and other ineligible units on a round by round basis with additional clerical services needed to process the forms.

- Additional telephone charges would be incurred for field supervisors calls to discuss special problems associated with the dwelling unit design that occur after Round 1

Table 3-3. Estimated Variable Costs Per Completed Interview to Complete Field Operations (in 1977 dollars)

Cost Item	NMCES Costs			Dwelling Unit Design
	Total Sample	Movers	Non-Movers	
Professional Labor	11.12	12.97	10.80	11.25
On-Site Labor	6.47	7.14	6.34	6.34
Off-Site Labor	4.65	5.83	4.46	4.91
Other Costs	37.72	45.28	36.41	37.84
Materials	0.47	0.52	0.47	0.48
Shipping/Communications	2.68	3.35	2.57	2.60
Travel	0.79	0.99	0.76	0.80
Report Printing	0.30	0.30	0.30	0.30
Interviewer Services	16.37	20.35	15.56	16.46
Interviewer Expenses	7.61	9.60	7.34	7.76
Respondent Incentives	3.68	3.68	3.68	3.68
Clerical Services	3.18	3.85	3.09	3.12
Printing	2.64	2.64	2.64	2.64
Total Technical Cost	48.84	58.25	47.21	49.09
Administrative Expense	4.33	5.22	4.19	4.37
Total Cost	53.17	63.47	51.40	53.46

(e.g., how to handle new construction or problems associated with the high rate of movement of individuals into and out of the survey).

- In validating interviewer work, Field Supervisors would need to travel after Round 1 (unlike NMCES) to verify that dwelling units listed as vacant were vacant, etc. This additional travel is listed under "travel". No additional travel is expected to be necessary for on-site personnel.

- Additional field interviewer services would be needed each round to check for new construction and to check the occupancy status of dwelling units that were unoccupied or whose residents previously refused to participate. More personal visits would be needed in telephone rounds to interview households whose occupants had moved since the previous round.

- Interviewer expenses would increase since more time would be spent in the field including additional travel costs incurred during telephone rounds.
- Since the Administrative Expense is costed as a surcharge on the Total Technical Cost, this line item would also change.

Based upon these results, it would appear that the unit cost of a completed interview using the dwelling unit design may be slightly larger than the unit cost incurred by the NMCES design or $53.46 versus $53.17. This is due to the fact that the cost savings due to confining interviewing activity within the original sample segments is offset by the additional screening and listing costs that must be incurred in each round for the alternative design and the additional personal interviewing that would be needed during telephone rounds.

From modeling the effects of nonresponse, attrition, and movement for the dwelling unit design, it would appear that a dwelling unit design not only will have a larger cost per completed interview, but that the total number of interviews might be larger (assuming the same number of dwelling units were selected). However, in costing the dwelling unit design, it was assumed that the total number of completed interviews would be the same as that of NMCES in order to facilitate a fair comparison.

3.3.2 Data Processing Cost Comparison

Most non-analytic data processing costs are dependent upon the number of documents to be processed and hence unit costs would not be affected by the sample design. However, some cost savings would result for the dwelling unit design from the fact that computerized monitoring of the day to day status of field operations (the control system) could use a simpler design with less records since the records could be dwelling unit level rather than individual level. Further, the dwelling unit design would not incur the costs that NMCES did for handling persons and families who move. The cost implications for various data processing activities are discussed below. In all comparisons, the same number of completed interviews is assumed for both designs.

Referring to Table 3-2, which presents NMCES costs per completed interviews, on-site data receipt costs would not change for the dwelling unit design since they depend only upon the number of documents handled. Data receipt includes manual log-in of documents received, editing and coding, and support for data collection activities in the field. Similarly, data entry cost for the dwelling unit design would not be affected since no additional documents would need to be keyed. Another category of costs appearing in Table 3-2 that would be unaffected by the sample design is the cost for coding of alpha-numeric variables.

Not included in the cost breakdowns given in Table 3-2 is the $21.17 per interview cost incurred by Abt Associates, Incorporated (AAI) for summary production and other data base tasks. NMCES produced a summary of all previously mentioned health care visits after each data collection round and mailed these summaries to the reporting units prior to the next interview. In producing the summaries, special problems resulted when individuals moved and either caused the creation of new reporting units or joined another reporting unit (when college students returned to their parents' home). Transfer of the data was often a problem when reporting units changed composition. In these cases, manual review by the summary coding staff was needed and then a manual system update by the programmer responsible for producing the summaries. The cost associated with resolving problems created by splits and merges are included in the $21.17 per interview cost for summary production and other data base tasks. Use of the dwelling unit design would result in an estimated savings of $0.18 per completed interview due to not having to follow individuals who move.

The remaining cost in Table 3-2, data processing labor and services cost, is associated with developing and maintaining the control system; producing control cards for each round; transmitting, reorganizing and editing the data; and other computer programming and services support for the survey. Some savings would occur in data processing for the dwelling unit design since the control system would be simpler and control card generation would be less complex.

The major source of savings for the dwelling unit survey would occur in the design and use of the control system. Since individuals would not have to be followed when they moved and formed new family units, only the reporting unit with its current family members would require monitoring on a round by round basis. These data would be used to update the control system with each round's current participants. Every inactive reporting unit (e.g., vacant dwellings) would require an event code to indicate its status. The number of records in the control system would be approximately

19,000 for the dwelling unit design as compared to 39,000 in the current NMCES since the unit being monitored would be the dwelling unit rather than the individual. In addition to the cost savings resulting from fewer records in the control system, the dwelling unit design would have other cost savings since the control system software needed by NMCES to handle movement and to set event codes for individuals within reporting units would not be required. The total savings due to having a simpler control system is estimated to be $1.04 per completed interview under the dwelling unit design.

Reporting units whose composition changed also resulted in problems as far as correctly producing control cards for the reporting units in the next round. These control cards contained household rosters and other data needed by the interviewer in conducting the interview for the round. The NMCES required manual handling for these reporting units changing composition that would not be needed in the dwelling unit design. Hard copy review and manual clean-up of the control card file was also required in NMCES, resulting in additional data processing labor costs that would not be incurred by the alternative design. In all, the dwelling unit design would produce control cards for $0.74 less per completed interview.

However, under the alternative design, a control card would be required for each dwelling unit the interviewer was expected to visit, whether that unit participated in the previous round or not. Thus, the number of control cards to be produced each round under the dwelling unit design would be the number of lines selected in the original sample instead of the number of households in the sample. In the NMCES, there were 63,209 control cards produced in Rounds 2 through 6 while the alternative design is projected to require the production of 90,375 control cards. The additional costs for computer services would total $0.49 per completed interview for the dwelling unit design.

The final costs to be considered under data processing are for transmitting, reorganizing, and editing the data. Since the instruments in NMCES were highly variable in terms of the types of records, record lengths, and number of records completed, special programs were developed to transmit the data from the data entry step and to reorganize the data into files of fixed length records. The assumption for costing the alternative dwelling unit design was that all NMCES instruments would be used and the same number of completed interviews would be obtained. Therefore, there would be no difference in costs for computer labor or services for the two designs for transmitting, reorganizing, and editing the data.

Balancing costs over all of the data processing tasks discussed as well as data receipt, data entry, data coding, and summary production, the alternative dwelling unit design as compared to NMCES would result in cost savings of $1.96 per completed interview and additional costs of $0.49 (due to the need for additional control cards), which together would produce a total data processing cost savings of $1.47 per completed interview had the dwelling unit design been used instead of a household design for NMCES.

3.3.3 Overall Cost Comparison

The majority of the costs for NMCES data collection and processing would have been unaffected by minor changes in the sample design such as those imposed by the dwelling unit design. Thus, there appears to be no substantial cost savings associated with the use of the alternative design. The field operations cost per completed interview for the dwelling unit design would have been $0.29 greater than that for the NMCES household design while the dwelling unit design would have produced an overall cost savings of $1.47 for data entry and processing. Thus, the dwelling unit design would result in a total cost savings of $1.18 per completed interview or a per completed interview total cost of $149.38 for the alternative relative to the NMCES total cost of $150.56.

3.4 Mobility Rates for Selected Subpopulations

Since the dwelling unit design does not include provisions for the follow-up of movers, many sample individuals would have missing data for one or more rounds if the dwelling unit design were used. To determine the impact this missing data might have on survey estimates of annual expenditures and utilization, mobility rates were computed for various subgroups of the population. These mobility rates were tabulated only for individuals interviewed in Round 1 and are unweighted estimates. A mover was detected by comparing the ZIP Code in Round 1 to the ZIP Codes for subsequent rounds and noting when the ZIP Code changed and hence the individual moved. Table 3-4 presents mobility rates for the domains defined by characteristics of the individual's Round 1 residence and Table 3-5 presents rates for domains defined by characteristics of the individual. Note that the rates given for each round are cumulative rates and reflect the percentage of individuals who moved before data

Table 3-4. Mobility Rates in Rounds 2-5 for Domains Defined by Geographic Characteristics

Domain	Round			
	2	3	4	5
Total	8.7	14.9	18.2	22.3
Region of the Country:				
Northeast	6.5	12.6	14.9	19.6
South	9.3	16.0	19.5	23.7
North Central	7.8	13.0	16.6	19.9
West	11.3	18.3	21.6	26.2
Size of Community:				
SMSA: Population of more than two million	8.0	14.3	17.6	21.7
SMSA: Central city of 500,000 or more	9.5	15.8	19.7	24.6
SMSA: Central city of less than 500,000	10.3	18.1	21.3	26.4
Non-SMSA: Less than 60 percent rural	8.1	13.6	16.1	19.5
Non-SMSA: 60 percent or more rural	7.4	12.4	15.3	18.3
Residence Characteristic:				
Inner city of SMSA central city	7.1	13.4	16.1	20.4
Remainder of SMSA central city	10.4	17.7	21.6	26.9
Urban area outside SMSA central city	8.6	14.6	19.1	24.7
Rural area outside SMSA central city	7.5	15.5	20.1	23.7
Urban area of a Non-SMSA	8.2	14.6	17.0	19.3
Rural area of a Non-SMSA	7.8	13.1	16.2	19.5

Table 3-5. Mobility Rates in Rounds 2-5 for Domains Defined by Personal Characteristics

Domain	Round 2	Round 3	Round 4	Round 5
Total	8.7	14.9	18.2	22.3
Race:				
White	8.8	15.0	18.3	22.3
Nonwhite	8.4	14.4	17.8	22.6
Sex:				
Male	8.7	14.9	18.2	22.4
Female	8.7	15.0	18.1	22.3
Age:				
0-4	9.6	16.6	20.0	24.8
5-14	7.3	13.7	16.8	20.8
15-24	12.3	19.7	23.6	29.0
25-34	9.2	17.3	20.8	26.0
35-44	7.7	13.5	16.8	20.7
45-54	7.1	11.9	15.1	17.6
55-64	7.2	11.8	14.0	17.0
65 or Over	7.4	11.7	14.6	17.9
Insurance:				
Medicare Part A	7.2	11.3	14.3	17.3
Medicare Part B	7.3	11.4	14.3	17.5
CHAMPUS or CHAMPVA	10.3	19.5	23.8	30.7
Other Public Assistance	7.5	14.4	17.8	24.5
Other Medical Assistance	11.2	16.9	19.0	20.7
Indian Health Service	12.5	12.5	12.5	17.9
Private Insurance	8.4	14.3	17.5	21.1
Uninsured	10.1	17.4	20.9	25.7
Income:				
Less Than $ 3,000	9.3	15.9	19.6	24.9
$ 3,000 - $ 4,999	10.5	17.5	22.1	26.5
$ 5,000 - $ 6,999	12.3	19.0	23.6	29.3
$ 7,000 - $ 9,999	8.8	17.1	20.3	24.9
$10,000 - $14,999	9.1	16.6	19.6	24.1
$15,000 - $24,999	8.0	13.8	17.0	21.3
$25,000 or More	7.3	13.1	16.8	20.4

collection activities for that round. Thus, the rates given for Round 5 are the percentages of individuals who moved in Round 5 or some previous round.

The mobility rates are rather similar for all domains with 22.3 percent of the total population moving by Round 5. The highest rates were found for the West (26.2%), standard metropolitan statistical areas (SMSAs) with central cities whose population was less than 500,000 (26.4%), and urban areas outside SMSA central cities (26.9%). The highest rates of movement were found for young adults (29.0% for individuals aged 15-24), for the $5,000 to $6,999 income group (29.3%), and for those individuals covered by CHAMPUS or CHAMPVA health insurance (30.7% movement was estimated using the unedited insurance variable).

3.5 Annual Utilization/Expenditure Distributions for Nonmovers Versus the Total Population

Many health care analyses require knowledge of the annual utilization and expenditures of individuals. When the dwelling unit design is used, a large proportion of the sample members will have data missing for one or more rounds. As discussed in the previous section, about one-fourth of the individuals interviewed during the first round of NMCES data collection had moved to a different ZIP Code area before the end of a year. Thus, when the dwelling unit design is used, a large proportion of the Round 1 individuals will move before the end of the year and hence have missing data for part of the year. Further, all those individuals who move into sample dwelling units after the first round of data collection will also have missing data for one or more rounds. In conducting any analysis that requires annual data, one may either perform the analysis on the data that are complete (e.g., the data from nonmovers) or one may impute for the missing data and then perform the analysis using the full data set. An imputation procedure should only be used that can produce estimates whose biases are substantially less than that of the estimates based upon the subset of complete records. Otherwise the cost of imputing for missing data will outweight any advantages with respect to bias reduction.

To effectively evaluate the dwelling unit design, the magnitude of the bias that will result from the incomplete data for movers must be known. To get a baseline estimate of the amount of this bias, the estimates for the nonmoving population were compared to the estimates for the total population. Using the NMCES data base, a mover can be detected by a change in the address from round to round. However, a complete comparison of addresses would have been difficult and time consuming to program since it would have involved comparing addresses containing alphabetic and numeric characters. Instead, for convenience and timeliness a mover was defined to be any individual whose ZIP Code at the end of the year (Round 5 of data collection) was different from that given in the first round of data collection. Thus, moves within the same ZIP Code area were not detected. Further, the ZIP Code was missing for five percent of the Round 1 sample individuals and hence their status as mover versus nonmover could not be determined. These 1,763 individuals with missing ZIP Codes for Round 1 include interviews that could not be completed until Round 2 and children that were born to sample members during the survey period. Since the movement status of these individuals without Round 1 ZIP Codes was unknown, all computations of bias in the rest of this chapter will be for known nonmovers versus the total population for which the Round 1 ZIP Code was known.

The first type of comparison is between the average annual utilization and expenditures for nonmovers versus that of the total population. Examining Table 3-6, one can see that little difference exists between the averages for the two populations. The absolute bias (i.e., the absolute value of the difference between the esti-

Table 3-6. Comparison of Average Annual Utilization/Expenditure Estimates for Nonmovers Versus the Total Population

Annual Average	Absolute Bias	Relative Bias (%)
Visits:		
Dental	0.000	0.00
Hospital	0.002	1.38
Ambulatory	0.075	1.51
Doctor in Hospital	0.010	0.42
Other Hospital	0.001	2.20
Total	0.079	1.14
Charges:		
Dental	0.27	0.98
Hospital	2.41	2.03
Ambulatory	1.54	1.90
Doctor in Hospital	2.34	8.22
Other Hospital	0.19	6.83
Total	1.39	0.54

mate for the known nonmovers versus the total population with known ZIP Codes) has a maximum value of 0.079 for visits and $2.41 for charges. The relative bias, defined as the ratio of the absolute bias to the estimate for the total population with known ZIP Code in Round 1, ranges from 0.00 percent to 2.20 percent for visits and from 0.54 percent to 8.22 percent for charges. These absolute biases are not large but may be important when compared to the standard error of the estimates; thus, the mean square error of the estimates may be substantially increased.

Another way of determining how different the medical care utilization and expenditures of nonmovers are from those of the entire population is to contrast the distribution of visits and charges for the two groups. The distribution pattern for nonmovers is again similar to that of the entire population with known ZIP Codes as may be seen from an examination of Table 3-7. The absolute bias for the ambulatory visit categories if data for nonmovers are used to make inferences about the total popuation ranges from 0.0038 percent to 0.7927 percent with relative biases from 0.20 percent to 3.37 percent. For hospital visits, the absolute bias ranges from 0.0012 percent to 0.1318 percent and the relative biases of the estimated proportions range from 0.15 percent to 10.21 percent. With respect to ambulatory and hospital charges, Table 3-8 reveals that the distribution for nonmovers is surprisingly similar to that of the entire population. If one were to use nonmovers to make inferences about the distribution of ambulatory charges for the total population, the absolute bias would range from a low of 0.0179

Table 3-7. Comparison of the Distribution of Ambulatory and Hospital Utilization for Nonmovers Versus the Total Population

Annual Utilization	Absolute Bias	Relative Bias (%)
Ambulatory Visits:		
0	0.7927	3.37
1-2	0.0542	0.20
3-4	0.1089	0.65
5-6	0.2352	2.41
7-8	0.0814	1.33
9-10	0.0810	1.96
11-14	0.0890	1.78
15-18	0.0825	3.11
19-22	0.0038	0.26
> 22	0.0567	1.75
Hospital Visits:		
0	0.1318	0.15
1	0.0808	0.90
2	0.0501	2.86
3	0.0016	0.35
4	0.0029	2.06
5	0.0043	10.21
> 5	0.0012	3.02

Table 3-8 Comparison of the Distribution of Ambulatory and Hospital Expenditures for Nonmovers Versus the Total Population

Annual Expenditures	Absolute Bias	Relative Bias (%)
Ambulatory Charges:		
$0	0.1749	0.45
$1 - $25	0.0980	0.59
$26 - $50	0.0261	0.21
$51 - $75	0.0288	0.36
$76 - $100	0.0381	0.70
$101 - $150	0.1311	1.94
$151 - $200	0.0231	0.62
$201 - $300	0.0514	1.35
$301 - $400	0.0660	3.90
> $400	0.0179	0.73
Hospital Charges:		
$0	0.1711	0.18
$1 - $99	0.0217	7.10
$100 - $299	0.0314	4.65
$300 - $499	0.0791	7.74
$500 - $699	0.0556	5.49
$700 - $899	0.0673	8.68
$900 - $1,099	0.0105	1.81
$1,100 - $1,599	0.0009	0.09
$1,600 - $2,099	0.0063	1.15

percent to 0.1749 percent with the maximum relative bias of an estimated proportion being 3.90 percent. Similar results can be noted for hospital charges with the maximum absolute bias found of 0.1711 percent. The relative biases for the distribution of hospital charges tends to be larger in general and ranges from 0.09 to 8.68 percent.

The comparison of average utilization/expenditure statistics and the distributions of these for nonmovers versus the entire population would indicate that medical care usage for the two populations is approximately the same. This result is confirmed by the finding that mobility rates are relatively similar for most of the major subdivisions of the population (see Section 3.3). However, both of these results were based upon defining a move as a change of ZIP Code and hence moves within the same ZIP Code area were not detected. Therefore, the results in this section should be interpreted with caution since the group referred to as nonmovers actually includes within ZIP Code movers. The results appear to indicate that movers may not be substantially different from nonmovers with respect to their medical care utilization and expenditures.

In some instances, the biases, though small in absolute value, are greater than the standard error of the estimates. As noted earlier, the bias that would result from using data for nonmovers to represent the entire population can have a serious impact when the bias is large with respect to the size of the standard error. The effect of bias when drawing inferences from survey data can be illustrated by examining the effect of large bias ratios (the bias divided by the standard error of the estimate). For each increase of 0.5 in the absolute value of the bias ratio, the probability will be approximately doubled that the corresponding interval fails to contain θ (Kish, 1965, pp. 566-571).

Thus, the bias detected in this examination of utilization/ expenditure statistics could potentially have an important impact on all confidence intervals and significance levels for subgroup comparisons. Further, the biases may not be as small for subdomains such as blacks, the young, or the poor and elderly. There is just reason to suggest that if complete annual data are of major importance for the statistical analysis of health care utilization and expenditures, then data from a dwelling unit design would not result in estimates of as high a quality as those obtained using data from a household design.

3.6 Concluding Remarks Regarding the Dwelling Unit Design

The results of the in-depth examination of the longitudinal unit design were surprising. At the start of this task, survey operations and data processing personnel felt that the dwelling unit design could have resulted in large cost savings had this design been used for NMCES instead of the household design. The dwelling unit design is a completely clustered design with no interviewing occurring outside of the original sample segments in any round. This is in contrast to NMCES, in which the original sample members were followed and hence location of the interviewing activity became more and more dispersed with subsequent rounds.

The cost savings of not having to follow individuals and/or families who moved was expected to be further enhanced by the fact that data processing could be simplified. Maintaining the control system and generating control cards for NMCES was complicated by round to round changes in the structure of the household. The difficulty caused by composition changes of the reporting units and the requirement that new reporting units be created to reflect these events would not occur for the dwelling unit design since only the current residents of the dwelling units would need to be monitored.

The detailed examination of costs for the dwelling unit design revealed that the cost savings for field operations due to not having to trace and follow individuals and/or families who move would be offset by the additional interviewer and field supervisor labor and expenses associated with screening and listing costs incurred after Round 1. The lower cost of following movers with respect to the screening and listing costs for the dwelling unit design can be attributed to the fact that the survey has a short time frame in which most movers do not travel very far from the original segments. In addition, the dwelling unit design requires personal interviewing, screening, and listing of in-migrants during telephone rounds whereas the household design does not incur these costs.

Most of the data processing costs are variable costs that depend upon the number of documents being processed. Thus, some cost savings may occur due to the simpler structure of the control system and the ease of generating control cards, but the savings would be small in relationship to the total cost for data processing. Assuming that the dwelling unit design resulted in the same number of completed interviews, a total savings of $1.18 would result from the use of the dwelling unit design.

At least half of the cost savings of the dwelling unit design might have to be spent if imputation were needed for individuals with data missing in one or more rounds. This imputation would be required for analyses that depend upon knowledge of annual data for individuals, since about 25 to 30 percent of the sample would have missing round data. Certain types of analyses such as the coding of visits with respect to episodes of illness would be severely hampered by the level of missing data associated with the dwelling unit design. All analyses that seek knowledge regarding the association of annual utilization/expenditures with other individual characteristics including regression or correlation analyses would also be seriously weakened. In light of the rather trivial cost savings projected for the dwelling unit design in relation to the total data collection and data processing budget for a NMCES-sized survey, the data analysis complications associated with the dwelling unit are very compelling and suggest that the household design is the best design for future health care utilization and expenditure surveys.

References

Bureau of the Census (1978). _The Current Population Survey: Design and Methodology_, Technical Paper 40, Washington, DC: U.S. Government Printing Office.

Cox, B., R. Folsom, B. Moser, and T. Virag (1979). Housing Unit Versus Household as the Sampling Unit, Report No. RTI/1725/01-01S, Contract No. 233-78-2102, National Center for Health Statistics, Hyattsville, Maryland.

Kish, L. (1965). Survey Sampling, New York: John Wiley and Sons.

National Center for Health Statistics (1965a). Comparison of Hospitalization Reporting in Three Survey Procedures. Vital and Health Statistics, Series 2, No. 8, PHS Publication No. 1000, Washington, DC: U.S. Government Printing Office.

National Center for Health Statistics (1965b). Reporting of Hospitalization in the Health Interview Survey. Vital and Health Statistics, PHS Publication No. 1000, Series 2, No. 6, Washington, DC: U.S. Government Printing Office.

National Center for Health Statistics (1972). Optimum Recall Period for Reporting Persons Injured in Motor Vehicle Accidents. Vital and Health Statistics, Series 2, No. 50, DHEW Publication No. (HSM) 72-1050, Washington, DC: U.S. Government Printing Office.

4
Data Collection Organization Effect

For laboratory measurements, researchers have long recognized that the organization making the measurements contributes to the total error of a given statistic, over and above the errors induced by the instrument itself or by the individual using the instrument. Variations in the average result of repeated measurements on the same material have been observed among laboratories using the same type of instrument and measurement protocol. This variation is due to a number of factors including age of the equipment, experience of the staff, and environmental conditions. For this reason, it is common to have independent measurements made by more than one laboratory to reduce the laboratory effect on the accuracy of study results.

Although the situation with respect to laboratory measurements could also extend to social surveys, little attention has been paid to this potential source of error in survey estimates. Despite a common set of survey conditions and measurement protocols, it can be demonstrated that repeated measurements with the same respondent and interviewer vary and that averages of measurements made of random subsamples of respondents, each assigned to a different interviewer at random, exhibit more variance than can be attributed to inherent variation (pure sampling error) among respondents. It is not inconceivable, then, that the organization responsible for carrying out a social survey may contribute to the realized survey error and that the extent of the contribution is the summation of a number of factors, including the overall experience and quality of the interviewing staff as well as the attention given by supervisory staff to survey protocol, to interviewer performance, and to quality control.

Conceptually, the total error in a given survey statistic includes a component due to the data collection organization. If the data collection organization effect varies among qualified organizations, a substantial contribution to the total error of survey estimates may occur, particularly in large surveys where sampling variation is small. With larger data collection organization effects, consideration would have to be given to using more than one data collection organization, with the decision based upon the magnitude of the organization error component relative to other components of error. Clearly, if the data collection organization effect is too small to be detected at the level of sensitivity possible with a given design and sample size, use of a single organization for data collection is appropriate.

Since NMCES used two survey organizations each independently selecting national samples and collecting data for the samples, a rare opportunity is afforded to estimate the magnitude of the data collection organization effect for a wide range of demographic and health care related variables. Strictly speaking, the data collection organization effect should be defined as the component of the total survey error for a given statistic that is due solely to the particular survey organization that collected the data. Factors that can contribute to observing detectable organization effects in NMCES estimates include any biases in the organizations' sample designs or systematic errors that may have been introduced during data collection by the organizations' interviewing and supervisory staffs.

In the study described in this chapter, national estimates of relevant demographic and health care parameters were separately produced for each of the two survey organizations to determine whether differences exist that suggest the presence of a data collection organization effect (Cohen and Horvitz, 1979; Cohen, 1982). This effect could be directly estimated and tested within the framework of an appropriate multivariate analysis that considers the complex nature of the survey design. Such an analysis was applied to a specific health care variable.

4.1 Related Research

Researchers' perceptions of whether different survey organizations produce similar measurements were examined by Goldfield, et al (1977). In the Goldfield study, data were collected separately by a government organization and a university center on public attitudes and knowledge about surveys, survey organizations, and issues of privacy and confidentiality. Of the respondents who believed a difference existed in the capability of alternative data collection organizations to obtain accurate information, 37 percent of the university center sample respondents indicated that the national government was most likely to get accurate reporting; 42 percent of the government sample respondents shared that position. In contrast, 29 percent of the university center respondents said that universities were most likely to obtain accurate reporting, whereas only 16 percent of the government respondents held the same opinion. The government organization also obtained a smaller nonresponse rate.

Smith (1978) examined whether different survey organizations produce similar measures of public opinion. A search was made for examples of different organizations asking identical questions at approximately the same time. Thirty-three examples were considered in which a question from the General Social Survey series overlapped with questions included in surveys conducted by (1) Gallup and the American Institute of Public Opinion, (2) the Survey Research Center of the University of Michigan, and (3) Roper. In ten out of thirty-three comparisons, significantly different responses were observed between organizations. Since these differences were concentrated among questions related to national spending, voting, and misanthropy, it was concluded that organization effects were not random in occurrence.

Further research by Smith (1982) tested for the existence of organization effects in a collaborative experiment between the

General Social Survey of the National Opinion Research Center, and the American National Election Study of the Center for Political Studies. Identical questions were asked by both surveys, related to the respondent's position and the position of the Federal Government on defense spending, minority assistance and social spending. It was not possible, however, to control for interviewer training, field procedures, instrument content, and field period. Significant organization effects were detected, with item nonresponse showing the largest and most systematic differences.

4.2 NMCES Organization Effects for Sample Design Parameters

The sample design for the National Medical Care Expenditure Survey was complicated in that it combined two independently drawn national samples of households, one by the Research Triangle Institute (RTI) and one by the National Opinion Research Center (NORC). Since the structure of both national area samples was similar, it was assumed that they were compatible and would allow the derivation of statistically equivalent, unbiased estimates of relevant health parameters. The NMCES data were collected by RTI and NORC for their respective national household samples with a common survey methodology. The two survey organizations used the same instruments, specifications, training procedures, field procedures, manuals, and quality control procedures in data collection. In addition, field supervisors for both organizations were trained together.

Even with a common survey methodology, organizational characteristics could result in differences (e.g. the personnel policies, management, and supervisory staffs). Further, the field staff of each organization hired its own interviewers, trained them, and supervised their performance. Thus, even though RTI and NORC operated under a common set of survey conditions with comparable sample designs, it would be possible for differences to exist between the data collected by the two organizations, over and above the differences due to sampling variations. These differences could occur due to nonsampling errors introduced by differences in structure and staff of the survey organizations.

The NMCES sampling design can be characterized as two independent replicates of similar four-stage samples of the civilian noninstitutionalized population of the United States. The first stage of both designs consists of primary sampling units which are counties, parts of counties, or groups of contiguous counties. These units were stratified by geographic location, degree of urbanization, and size for both organizations and also by percentage black for NORC. The second stage consists of secondary sampling units, which are generally Census-defined block groups or enumeration districts in both designs. Smaller area segments constitute the third stage of the designs, with the selection of one segment per secondary unit. Selection in each of the first three stages was with probability proportional to size. At the fourth stage of sampling, dwelling units were selected from a machine-readable frame by a systematic sampling method, yielding a national probability sample of approximately 13,500 dwelling units (Cohen and Kalsbeek, 1978, 1981).

Reporting units were established to account for the fact that dwelling units may contain unrelated persons, whereas for analysis and data collection purposes, groups of related individuals (i.e., families) had to be identified. The unit for which a questionnaire booklet was completed, a _reporting unit_ was defined as a family unit living within the same dwelling unit. Data were collected from all individuals within the initial sample dwelling units in six rounds of interviews conducted throughout 1977 and during the first half of 1978. College students living away from home were included when their parents' residence was sampled and were assigned to a separate reporting unit to facilitate interviewing. In each round, new reporting units were constructed to reflect changes in the composition of sample individuals' families.

A summary of design differences between the two survey organizations is presented in Table 4-1. The RTI half-sample of the NMCES consisted of 59 primary sampling units yielding 644 secondary sampling units and 644 segments, an average of 10.9 segments per primary sampling unit. A total of 19,598 sample individuals responded in the RTI half-sample of the NMCES. The NORC half-sample consisted of 76 primary sampling units which yielded 646 secondary sampling units and 646 segments, an average of 8.5 segments per primary sampling unit. The NORC half-sample produced a total of 19,220 responding individuals. When the two half-samples are combined, 27 primary sampling units are common to both designs, resulting in more segments per primary sampling unit for the combined design than for the two individual half-samples.

Table 4-1. Summary of NMCES Sampling Data

Characteristic	Organization NORC	RTI	Combined Sample
Primary Sampling Units	76	59	108[1]
Segments	646	644	1,290
Average Segments per Primary Sampling Unit	8.5	10.9	11.9[2]
Reporting Units	7,562	7,673	15,235
Reporting Units Providing Data for All Rounds of Data Collection	6,181	6,304	12,485
Reporting Units Ineligible for One or More Rounds of Data Collection	69	48	117
Number of Individuals	19,220	19,598	38,818

[1]Twenty-seven primary sampling units were common to the two samples.

[2]The average number of segments per primary sampling unit is greater for the combined sample due to the 27 primary sampling units that belong to both general purpose household samples.

Another area of potential difference between organizations is in the extent to which they can obtain complete responses from sample members. Measuring overall response for the NMCES is difficult since response was monitored on a reporting unit basis and reporting units changed with time. New reporting units were created after Round 1 to account for family units formed by sample individuals leaving home and for family members going away to college. Similarly, reporting units were dissolved when all sample members became ineligible (i.e., died, left the country, joined the military, or entered an institution) and when college students living away from home rejoined their families during vacations. Hence, when an interview was not completed in all data collection rounds for a reporting unit, data may or may not be missing for the members of the reporting units. Furthermore, when an interview could not be obtained in a round (e.g., family were never at home, could not be located, or refused), NMCES attempted to collect data for the missing time period in the next round. When these events are considered, the NMCES achieved a 91 percent response rate in Round 1 and obtained complete 1977 data for approximately 89 percent of the Round 1 participants, resulting in an overall complete data rate of 81 percent. Unfortunately, these rates are not available for RTI and NORC separately.

To determine whether there appears to be a difference between the overall performance of RTI and NORC, two related rates (not complete data rates) can be calculated based upon Table 4-1. The percent of all reporting units for which data were obtained for all rounds is 81.7 for NORC and 82.2 for RTI. When reporting units that were not eligible for the full year are removed, the percent of reporting units providing data for all rounds is 82.5 for NORC and 82.7 for RTI. Based upon these two rates, there would seem to be no appreciable difference in the performance of the two organizations with respect to obtaining response.

Estimation in the NMCES requires the use of analysis weights that reflect the variation in selection probabilities for observational units, in addition to nonresponse and post-stratification adjustments computed separately for each survey organization. The NMCES post-stratification adjustment forced the nonresponse-adjusted population estimates for each survey organization to Census Bureau projections of the 1977 population within weighting classes defined by age (0-4, 5-14, 15-24, 25-34, 35-44, 45-54, 55-64, 65 and older), race (white, nonwhite), and sex (male, female).

Since the RTI and NORC sample designs allow for unbiased national estimation, the population size estimate for the two samples should be approximately equal prior to post-stratification for each of the 32 weighting classes. To verify this, Studentised differences between the estimated 1977 population counts for the two samples were tested to determine if they were significantly different from zero.* Each organization's estimates were derived independently from analysis weights adjusted only for nonresponse. The null hypotheses of interest were the equivalence of estimated population totals across samples. The results of these tests, which are shown in Table 4-2, reveal that only the estimated total white females less than five years old differs significantly between samples at an α level of 0.05 (Cohen and Horvitz, 1979). With 32 comparisons being made at the 95 percent confidence level, chance will result in an average of 1.6 comparisons being rejected when the

---

*The Studentised difference is the difference between the estimates divided by the standard error of the difference. The Studentised difference is asymptotically normal, and has mean zero and variance one when the two estimates are equal.

two sets of sample estimates are equivalent. Consequently, the RTI and NORC estimated population totals for the 32 weighting classes can be regarded as equivalent.

Another important consideration is whether these estimated population totals differ significantly from the independent and more precise set of estimates provided by Census Bureau population projections for 1977. In NMCES, the sampling weights for each survey organization were adjusted to the Census population projections. The post-stratification adjustments made for each of the 32 weighting classes are also presented in Table 4-2. Generally, the estimated totals derived for NORC from the analysis weights (adjusted only for nonresponse) produced underestimates compared to the Census projections, whereas the RTI analysis weights more often yielded overestimates than underestimates.

To determine whether the analysis weights from each survey organization were significantly modified when forced to the Census totals, the population estimates derived from each survey organizations were individually tested for difference from the Census projection using the Studentised statistic and an α level of 0.05.* This test between the estimated population totals and the Census projections translates to testing the post-stratification adjustments to determine whether they are significantly different from one. For the RTI sample, only two of the 32 adjustments appeared to

---

*Here, the Studentised statistic is the difference between the organization's population estimate and the Census projection, divided by the standard error of the organization's population estimate.

Table 4-2. Post-Stratification Adjustments And Estimated Population Totals Prior to Post-Stratification By Contractor and Weighting Class

Weighting Class			Post-Stratification Adjustment		Population Estimate		Studentised Difference Between Totals
Race	Sex	Age	NORC	RTI	NORC	RTI	
White	Male	0-4	1.1442*	1.0956	5,629,275	5,878,735	-0.6212
		5-14	1.0110	1.0018	15,276,731	15,416,786	-0.1288
		15-24	1.0051	0.9386	16,660,552	17,850,972	-1.1022
		25-34	1.0630	1.1418*	12,981,525	12,085,633	1.1391
		35-44	1.0215	0.9857	9,614,972	9,964,425	-0.6032
		45-54	1.0673	1.0480	9,347,152	9,518,595	-0.2869
		55-64	0.8950	0.9311	9,636,427	9,262,666	0.6333
		65+	0.9121	0.8638	9,092,561	9,600,417	-0.6027
	Female	0-4	1.0448	1.2305*	5,865,140	4,979,919	2.0824†
		5-14	1.0205	1.0056	14,481,071	14,696,414	-0.2387
		15-24	1.0116	1.0417	16,932,589	16,442,269	0.5480
		25-34	1.1066*	1.0880	12,899,161	13,119,961	-0.3186
		35-44	1.0028	0.9917	10,348,284	10,464,052	-0.1605
		45-54	0.9882	1.0144	10,721,804	10,444,648	0.5101
		55-64	0.9230	0.9991	10,395,302	9,603,478	1.2386
		65+	0.8913*	0.8783	13,317,051	13,514,507	-0.2103

Non-White	Male	0-4	1.0959	0.8083	1,248,569	1,692,836	-1.5822
		5-14	1.0022	0.9425	3,031,203	3,223,177	-0.3525
		15-24	0.9657	0.9390	2,827,043	2,907,411	-0.1422
		25-34	1.1685	0.7803	1,547,806	2,317,177	-1.4821
		35-44	0.9874	0.9077	1,292,328	1,405,773	-0.4603
		45-54	1.0980	1.0438	1,097,537	1,154,517	-0.2612
		55-64	1.1342	0.9187	787,811	972,655	-0.9841
		65+	1.2971	0.7539	696,438	1,198,186	-1.6789
	Female	0-4	0.9618	0.7613	1,389,957	1,755,990	-1.1426
		5-14	1.0973	0.9403	2,741,971	3,199,935	-0.8029
		15-24	1.0096	0.8867	3,057,465	3,481,151	-0.7432
		25-34	1.0531	0.8355	2,173,212	2,739,121	-1.1737
		35-44	1.1233	0.9620	1,452,075	1,695,580	-0.8427
		45-54	1.0340	1.0639	1,367,540	1,329,076	0.1548
		55-64	1.0370	0.8000	1,015,525	1,316,274	-1.0299
		65+	1.1862	0.8643	1,011,429	1,388,035	-1.0744

*Indicates an adjustment factor that is significantly different from one at the 0.05 confidence level.

†Indicates a significant difference between the two population estimates at the 0.05 confidence level.

significantly alter the values of the nonresponse adjusted analysis weights: the adjustments for white males 25 to 34 years old and the adjustment for white females less than five years old. This is approximately equivalent to the number of differences that could be expected to occur by chance, giving additional assurance of the reliability of the preliminary sampling weights. The NORC sample yielded three statistically significant adjustments: the adjustment for white males under age five and the adjustments for white females aged 25 to 34 and 65 years or older. This was slightly above the 1.6 differences that could be expected to occur by chance when testing at an α level of 0.05.

4.3 A Comparison Of Demographic And Health Care Measures Between Survey Organizations

A major concern of this study was the capability of the two survey organizations to yield statistically equivalent estimates of demographic and health care parameters relevant to the NMCES. The demographic measures under investigation included region, size of city, family income, marital status, health status, and years of school completed. Estimates of the national distributions for these demographic variables were derived for the two survey organizations and are presented in Table 4-3. To determine whether the distributions were equivalent across survey organizations, a test of homogeneity appropriate for data from a complex survey design was applied.

Estimated proportions for each demographic distribution and their associated variance-covariance matrices were calculated for the two survey organizations and weighted least squares methodology was used to test for equivalence in parameter estimates across survey organizations.* To account for the effects of clustering and stratification induced by the complex sample designs, variances and covariances of sample estimates were derived using the Taylor Series linearization method (Woodruff, 1971 and Shah, 1981).

Testing at the 0.05 confidence level, no significant differences in the demographic distributions were evident across samples. The tests for equivalence of the estimated distributions for the region, size of city, family income, and marital status measures resulted in nonsignificant test statistics, indicating consistency in parameter estimates across survey organizations. The tests of the distribution of health status and years of school completed also supported the hypothesis of statistical equivalence between the survey organizations, though not to the same degree. This was undoubtedly a result of the larger levels of item nonresponse encountered by NORC for these particular measures.

The following health care indices were also investigated: (1) the average number of medical provider visits per person, (2) the average length of time (in days) a person waited to see a medical provider after making an appointment, (3) the proportion of visits for which patients were dissatisfied with the wait in the doctor's office, and (4) the proportion of the U.S. population ever on Medi-

---

*This methodology has been described by Grizzle, et al (1969) and implemented in the analysis of data from complex surveys by Koch, et al (1975) and Freeman, et al (1976). The methodology is described in greater detail in Chapter 12.

Table 4-3. A Comparison of Demographic Measures Between Survey Organizations

Demographic Measure	NORC (%)	RTI (%)
Census Region:		
Northeast	21.2	23.2
North Central	27.6	27.1
South	31.7	31.9
West	19.5	17.8
Q = 0.496 (3 degrees of freedom), P = 0.920*		
Size of City:		
SMSA: 16 largest SMSAs	24.1	27.5
SMSA: 500,000+ population but not 16 largest	26.4	24.7
SMSA: less than 500,000 population	20.6	15.7
Not SMSA: less than 60% rural	17.3	19.7
Not SMSA: 60% or more rural	11.6	12.4
Q = 1.723 (4 degrees of freedom), P = 0.787*		
Family Income:		
Under $8,000	22.1	20.7
$8,000 to $17,999	34.2	34.5
$18,000 and over	43.7	44.8
Q = 1.561 (2 degrees of freedom), P = 0.458*		
Marital Status:		
Under 17 years of age	28.2	28.3
Never married	13.4	13.3
Married	43.8	44.7
Widowed	5.6	5.2
Separated	1.7	2.2
Divorced	3.7	3.1
Unknown	3.6	3.2
Q = 10.814 (6 degrees of freedom), P = 0.094*		

Table 4-3. (continued)

Demographic Measure	(%) NORC	(%) RTI
Health Status:		
Excellent	43.0	45.6
Good	36.5	38.0
Fair	10.4	10.3
Poor	3.6	3.1
Unknown	6.5	3.0
Q = 8.468 (4 degrees of freedom), P = 0.076*		
Years of School Completed:		
0 to 8 years	10.9	10.8
9 to 11 years	13.1	12.7
12 years	23.0	25.0
13 to 15 years	10.6	10.7
16 or more years	8.6	9.2
Under 17 years of age	25.7	26.7
Unknown	8.1	4.9
Q = 11.863 (6 degrees of freedom), P = 0.065*		

*Testing for homogeneity in distributions across organizations and determining the P-value for the Wald statistic, Q, by comparison to a χ^2 distribution with the appropriate degrees of freedom. The P-value is the probability of obtaining a value of the test statistic as extreme or more extreme than the one actually computed, under the null hypothesis of no difference between the two distributions. When $P \leq 0.05$, the null hypothesis is rejected when testing at the 0.05 level of significance.

caid in 1977. Estimates of these health care parameters are shown in Table 4-4. To test for the equivalence in health care estimates across organizations, Studentised differences were used. All the comparisons showed a close correspondence between the estimates from the two survey organizations, and no significant differences were noted when testing at an α level of 0.05.

Table 4-4. A Comparison Of Health Care Measures Between Survey Organizations

Health Measure	Estimate and Standard Error		Studentised Difference	P-Value*
	NORC	RTI		
Mean Number of Medical Provider Visits	5.19 (0.10)	4.98 (0.12)	1.36	0.174
Mean Waiting Time in Days to See Medical Provider	7.74 (0.36)	7.34 (0.25)	0.90	0.368
Proportion of Visits Where Patients are Dissatisfied with Wait in Doctor's Office	0.139 (0.005)	0.136 (0.004)	0.52	0.623
Proportion of U.S. Population Ever on Medicaid	0.105 (0.007)	0.106 (0.006)	-0.16	0.873

*The P-Value is the probability of obtaining a value of the test statistic as extreme or more extreme for the two-sided test than the one actually computed, under the null hypothesis of no difference between the two organizations' estimates.

4.4 Estimated Data Collection Organization Effect

The results of the comparison of demographic and health care measures support the notion of a nonsignificant organization effect. Analyses were also made to test for organization effects by domain since organizational domain differences may be cancelled out or masked in total population analyses. The mean number of medical provider visits was chosen for the detailed domain analyses since it exhibited the greatest dissimilarity across organizations.

Control variables were introduced to define the domains and examine interaction effects with organization. Family income, residence (metropolitan, nonmetropolitan), and race were selected as control variables as a consequence of their observed association with the utilization measure being examined. Twenty-four distinct domains were determined by a cross-classification of these demographic factors with organization. Once the domain estimates and associated variance-covariance matrix were calculated, weighted least squares methodology was again used to test for a data collection organization effect.

The analysis consisted of the specification of an underlying linear model for the utilization measure that included main effects and first and second order interactions for organization and the control variables. A Wald statistic for goodness of fit was used on the residual sum of squares for the model to determine its appropriateness. Testing at an α level of 0.05, Wald statistics were used to determine whether the model effects were significant. The approach allows for an assessment of the statistical significance of different sources of variance in the data through a partition of model components. In this manner, it is possible to test for an organization effect in addition to the interaction effects of the control variables with organization.

The test for data collection organization effect considered the variation between domain estimates of the mean number of medical provider visits. A nonsignificant goodness of fit test statistic was observed when testing at an α level of 0.05, suggesting the specified model was reasonable (Table 4-5). Observed differences in domain estimates were completely due to the influence of race, family income and residence. With these factors controlled, differ-

Table 4-5. Test Results for Organizational Effect on Estimation of Mean Number of Medical Provider Visits

Test of Hypothesis*	Degrees of Freedom	Q	P-Value
Survey Organization	1	0.570	0.450
Combined Main Effects of Family Income, Residence and Race	4	124.823	<0.001
Interactions of Control Variables with Organization:			
First Order	4	1.856	0.762
Second Order	5	7.462	0.189
Interactions of Control Variables:			
First Order	5	7.636	0.178
Second Order	2	6.499	0.039
Goodness of Fit	2	0.886	0.642

Estimated Organization Effect = 0.088, Standard Error = 0.116

*The model took the form $y = X\beta$ where y was a 24 by 1 vector of domain estimates, X was a 24 by 22 prespecified design matrix, and β was a 22 by 1 vector of effects.

ences in utilization estimates for the two survey organizations were found to be nonsignificant. In addition, no significant first (two-factor) and second order (three-factor) interaction effects of the control variables with organization were evident.

4.5 Concluding Remarks

In this study, an assessment was made of whether two survey organizations, with similar national probability designs and similar survey protocols, could produce statistically equivalent national estimates of relevant demographic and health care parameters. The findings supported the notion of statistical equivalence on three levels:

- the capacity to yield statistically equivalent, age-race-sex national population estimates that were also consistent with Census Bureau projections,
- the ability to provide statistically equivalent, distributional estimates of demographic measures which include region, size of city, family income, marital status, health status, and years of school completed, and
- the consistent observation of a nonsignificant data collection organization effect when testing for differences in estimates of alternative health indices.

The latter finding was confirmed by the more detailed domain analysis for the utilization measure. Differences observed between domain estimates were explained by residence, race, or family income differentials, and no significant interaction effects between organization and the control variables were detected.

The results of this chapter should not be construed to indicate that organizational effects do not exist. Rather, the results of this study indicate that two similarly qualified organizations can collect comparable survey data when a common survey methodology is implemented. The organizations that conducted the NMCES were not selected randomly from all possible organizations but purposively selected based upon cost and quality considerations. If a greater mix of organizations in terms of organizational size, staff experience, and overall qualifications had been used to collect NMCES data, differences between organizations might have been detected.

Considerations other than organization effect may lead to the use of more than one data collection firm for very large surveys.

For instance, the NMCES had stringent time constraints on data collection and processing, many hinging upon the fact that NMCES data collection rounds were approximately three months apart. In the time between Round 2 and Round 3, for instance, the data for Round 2 had to be collected, keyed, edited, coded, and entered into the data base. The data base then had to be used to generate a cumulative summary of the household's health care utilization and expenditures, which had to be mailed to the household and the interviewer prior to Round 3. Data collection and processing for approximately 40,000 sample individuals in such short periods of time is beyond the capacity of most data collection organizations.

Another advantage of involving more than one contractor relates to the quality of the work done. An organization's access to experienced interviewing and supervisory staff is related to the volume of work they normally do. Similar remarks may also be made about the in-house staff needed to monitor data collection, to edit and key the data, and to produce the final data base. Merging the resources of more than one organization enlarges the pool of experienced staff which can be assigned to the task.

The disadvantage of more than one contractor lies in the unavoidable duplication of effort. For example, in the NMCES combined sample of 135 primary sampling units, only 108 of the units were unique. Because of the sample duplication, two sets of field staff were used in 27 primary sampling units. In addition, each organization had to incur fixed costs associated with sampling, data collection, and data processing. In a design study based upon the 1980 NMCUES, Cox, et al (1983, pp. 2-16 to 2-21) estimate that the cost penalty associated with duplication of effort may be substantial. This cost penalty associated with sample duplication together with the lack of organizational differences between qualified organizations found in this study suggest that consideration of more than one organization for data collection should be restricted to situations in which the capability of one organization to do the study is in question.

References

Cohen, S. B., and D. G. Horvitz (1979). Estimated Data Collection Organization Effect in the National Medical Care Expenditure Survey. Paper presented at the 1979 meetings of the American Public Health Association and available from the National Center for Health Services Research, Rockville, Maryland.

Cohen, S. B., and W. D. Kalsbeek (1978). Some Statistical Implications on Analysis of the Design of the National Medical Care Expenditure Survey. Paper presented at the 1978 meetings of the American Public Health Association and available from the National Center for Health Services Research, Rockville, Maryland.

Cohen, S. B. (1982). Estimated Data Collection Organization Effect in the National Medical Care Expenditure Survey. _American Statistician_, 36, 337-341.

Cohen, S. B., and W. D. Kalsbeek (1981). _NMCES Estimation and Sampling Variances in the Household Survey_, NHCES Instruments and Procedures 2, DHHS Publication No. (PHS) 81-3281, Washington, DC: U.S. Government Printing Office.

Cox, B. G., R. E. Folsom, T. G. Virag, and W. F. Refior (1983). _Design Alternatives for Integrating the NMCUES with the NHIS_, RTI Final Report No. RTI/1900/40-01F, National Center for Health Statistics, Hyattsville, Maryland and the Health Care Financing Administration, Baltimore, Maryland under Contract No. HRA-233-79-2032.

Freeman, D. H. Jr., J. L. Freeman, D. B. Brock, and G. G. Koch (1976). Strategies in the Multivariate Analysis of Data from Complex Surveys II: An Application to the United States National Health Interview Survey. International Statistical Review, 44, 317-330.

Goldfield, E. D., A. G. Turner, C. D. Cowan, and J. C. Scott (1977). Privacy and Confidentiality as Factors in Survey Response. Proceedings of the American Statistical Association, Social Statistics Section, 219-229.

Grizzle, J. E., C. F. Starmer, and G. G. Koch (1969). Analysis of Categorical Data by Linear Models. Biometrics, 25, 489-504.

Koch, G. G., D. H. Freeman Jr., and J. L. Freeman (1975). Strategies in the Multivariate Analysis of Data from Complex Surveys. International Statistical Review, 43, 59-78.

Shah, B. V. (1981). SESUDAAN: Standard Errors Program for Computing of Standardized Rates From Sample Survey Data. RTI Report No. RTI/5250/00-01S, Research Triangle Institute, Research Triangle Park, North Carolina.

Smith, T. W. (1978). In Search of House Effects: A Comparison of Responses to Various Questions by Different Survey Organizations. Public Opinion Quarterly, 42, 443-463.

Smith, T. W. (1982). House Effects and the Reproducibility of Survey Measurements: A Comparison of the 1980 GSS and the 1980 American National Election Study. Public Opinion Quarterly, 46, 54-68.

Woodruff, R. S. (1971). A Simple Method for Approximating the Variance of a Complicated Estimate. Journal of the American Statistical Association, 66, 411-414.

5

A Demonstration of Optimal Allocation

As discussed in Chapter 2, stratification is frequently used in sample selection to increase the precision of survey estimates and to provide greater control over the distribution of the sample. Except when oversampling of certain population subgroups is needed, the allocation of the sample to strata is usually made proportional to the size of the strata. However, when data collection costs and variances differ among strata, optimal allocation of the sample to strata should be considered since it may result in substantial cost savings. Optimal allocations are usually designed to minimize total survey cost subject to specified variance constraints. These variance constraints control the precision of key survey statistics for the total population and for important reporting domains.

To optimally allocate the sample among strata, cost and variance models are needed. In practice, the following cost model is usually adequate for most sample designs with H strata:

$$C = C_o + \sum_{h=1}^{H} C_h n_h \qquad (5\text{-}1)$$

where

C is the total survey cost,

C_o is the fixed administrative cost of the survey,

C_h is the cost of surveying a unit from the h-th stratum, and

n_h is the sample size for the h-th stratum.

A corresponding variance model for survey estimates is:

$$V = \sum_{h=1}^{H} V_h / n_h \qquad (5\text{-}2)$$

where

V is the variance of the estimate, and

V_h is the variance component associated with sampling from the h-th stratum.

These cost and variance models illustrate that as the sample size for each stratum increases the variance decreases while the total cost of the survey increases.

The problem of determining the optimum sample size for the H strata involves specifying the maximum variance (V^*) allowed for a survey estimate. This problem may be represented mathematically as finding the set of within stratum sample sizes n_h that minimize the total survey cost C subject to the achieved variance being less than the maximum ($V \leq V^*$) and all strata being represented in the sample ($n_h > 0$ for all h). For this single variance constraint problem, the optimal allocation to stratum h is (Hansen, Hurwitz, and Madow; 1953; pp. 220-223):

$$n_h = [V_h/C_h]^{\frac{1}{2}} \sum_{h=1}^{H} [V_h C_h]^{\frac{1}{2}} / V^* \quad (5\text{-}3)$$

Under optimal allocation, the within stratum sampling rate will tend to increase as the variance increases and data collection unit costs decrease.

In practice, very few surveys are conducted with the sole objective of obtaining an estimate for a single population parameter. Sample allocation based upon the single variance constraint model described above would involve prioritizing different estimates and selecting the most important one for use in the optimization. A more appealing strategy is to consider several estimates simultaneously where the estimates are chosen by classifying the survey estimates according to their variance properties and selecting a typical variance model from each class. In contrast to the single constraint case, optimization for multiple variance constraints does not have a closed form solution; Cochran (1977, 119-123) reviews a number of approaches to obtain exact solutions for these problems.

Practicing statisticians commonly apply approximate techniques or develop compromise designs rather than use these rigorous methods which may lead to reduced survey costs but are more difficult to apply. This chapter presents a demonstration of design optimization where the optimal solution was obtained using a survey design optimization approach developed by Chromy and described in Folsom, Williams, and Chromy (1980). Chromy's optimization algorithm is an iterative approach that provides an optimal solution when the convergence criteria are met. In this investigation, total survey cost was minimized subject to fixed variance constraints.

5.1 The Allocation Problem

Current planning for population based surveys conducted by the National Center for Health Statistics (NCHS) assumes that the data systems can be integrated to save on data collection costs, to reduce respondent burden, and to increase the utility of the resultant data. As a part of a larger NCHS effort to evaluate the advantages of an integrated data system, alternative designs were examined for integrating the National Medical Care Utilization and Expenditure Survey (NMCUES) with the larger National Health Interview Survey (NHIS). This chapter summarizes the results of the investigation to assess the possibility of linking the two studies (Cox et al, 1983).

As a baseline for comparison, specifications were developed for an unlinked NMCUES design. Selected independently of the NHIS, this unlinked design would result in a stratified, clustered area sample similar to that of the NMCUES conducted in 1980, hereafter referred to as the "1980 NMCUES." For the sake of flexibility for NCHS planning, two sample sizes were used: 6,000 versus 10,000 responding households. The 6,000 household design is roughly equivalent in size to the 1980 NMCUES. The 10,000 household design was added so that NCHS could evaluate the improved precision for smaller domains versus the increased survey cost when the larger sample was used.

Linkage of the NMCUES to the NHIS implies that the NHIS sample selections can be used to form the frame for the NMCUES. Unlike the area frame used for previous surveys, the NHIS-based frame will contain names, addresses, and individual and family level characteristics. This information could be used in sample selection and data collection to reduce costs and improve data quality. The elements of the NHIS-based frame could be either the NHIS sample dwelling units or the NHIS sample households. For simplicity of presentation, this chapter will be restricted to the linked household design in which the frame is composed of NHIS sample households.

The linked household design would select a subsample of the NHIS sample households for inclusion in the NMCUES. The individuals within the subsampled households would be interviewed in Round 1 regardless of whether or not they lived in the clustered NHIS sample dwelling units. Rounds 2 through 5 data collection would follow the same rules as the unlinked design. Two sample sizes were again used in developing sample designs; these sample sizes were determined as the sizes required to yield the same precision as the unlinked design with 6,000 and 10,000 responding households. Optimal allocation was used to determine the number of PSUs and segments from the NHIS to include in the sample.

These two designs are self-weighting; that is, all sample individuals are selected with the same probability. In many ways, this eliminates the chief advantage of linkage with the NHIS. With knowledge of individual characteristics available for NHIS sample respondents, added precision can be obtained for small domains without proportionally increasing the size of the total NMCUES sample. To evaluate this feature of NHIS linkage, a third and final design type was investigated. This design is an optimally-allocated, linked-household design in which precision constraints were set for the total population and the Medicaid population based upon those achieved by the unlinked design.

An important result of this investigation is that there appears to be little relative gain from linkage when the final design has to be self-weighting. This result is reasonable when one considers that the principal gain from the linked self-weighting design is in the elimination of costs associated with counting and listing. Since the NMCUES mode of interview across rounds was not altered in this investigation (the first two rounds and the last round use personal interviews and the other two rounds use telephone interviews), there was little gain derived from the knowledge of names, addresses and telephone numbers for NHIS sample individuals. The pay-off came with the optimally allocated design that used characteristics of the NHIS respondents to oversample heavy users of health care services and to increase the precision for small domains without proportionally increasing the size of the total sample.

5.2. The Unlinked NMCUES Design

The unlinked NMCUES design that was studied in this investigation was patterned after the design used for the 1980 NMCUES. Specifically, an area sampling approach was assumed which would result in a self-weighting design in the sense that each sample individual would be selected with equal probability. The sample sizes required to yield 6,000 and 10,000 responding households had to be determined as well as the survey costs associated with these designs. The variances achieved by the unweighted, unlinked NMCUES design were then modeled for use in sample size determination for the remaining designs.

5.2.1 Definition of the Unlinked Design

The unlinked sample design would be a stratified, multi-stage area probability design in which each sample dwelling unit is selected with equal probability. The first stage sample would consist of primary sampling units (PSUs) which are counties, parts of counties, or groups of continguous counties. The second stage sample would consist of secondary sampling units (SSUs) which are Census enumeration districts (EDs) or block groups (BGs). Smaller area segments would constitute the third stage. All of the dwelling units within the sampled segments would be listed. During the fourth stage of sampling, dwelling units within each sample segment would be designated for inclusion in the NMCUES sample. All civilian, noninstitutionalized individuals residing in the sampled dwelling units in the initial round of data collection would be included in the survey. To facilitate family-level analyses, single college students in the 17 to 22 age range would be linked to their parents' residence and included in the survey only if their parents' residence was selected.

5.2.2 Sample Size Determination

Two sets of sample sizes were required for the unlinked NMCUES design: a sample size sufficient to yield 6,000 responding households and a sample size sufficient to yield 10,000 responding households. In order to obtain these sizes, a precise definition was needed for what constituted a "responding household" since households change over the data collection year. The decision was made to use responding Round 1 households and to describe the sample sizes needed to yield 6,000 and 10,000 responding Round 1 households. These Round 1 households are the Round 1 reporting units (RUs) after college student RUs are linked back to their parent RUs. The unit for which a questionnaire booklet is completed, a reporting unit is composed of individuals related by blood, marriage, or adoption who live in the same dwelling unit. Because data collection costs are for reporting units and rounds, sample sizes had to be developed in terms of these units.

The first step in this process was to model the 1980 NMCUES experience. The modeling began with the set of control system records generated by responding Round 1 households. (In the 1980 NMCUES, a Round 1 household was defined to be responding if it was linked to an RU that completed an interview in any of the five data collection rounds.) The 1980 NMCUES contained 6,269 responding Round 1 households. These responding Round 1 households generated 6,603 completed RU interviews in Round 1; 6,519 completed RU interviews in Round 2; 6,528 completed RU interviews in Round 3; 4,559 completed RU interviews in Round 4; and 6,561 completed RU interviews in Round 5. There tended to be more RU interviews than there were responding Round 1 households because households containing college students required more than one RU assignment to handle the different addresses for data collection. These interviews occurred over 108 unique PSUs and 809 segments.

5.2.3 Variance Modeling for the Unlinked Design

As a baseline for comparison of the unlinked with the linked designs, the decision was made to fix the precision of the linked designs for selected key statistics and key domains to that of the unlinked design, and then to compare the designs with respect to sample sizes and costs. The domains of interest for this demonstration were the total population and Medicaid recipients. The statistics of interest were as follows:

- average number of hospital visits,
- average number of facility visits,
- average number of office visits,
- average annual expenditure for hospital visits,
- average annual expenditure for facility visits,
- average annual expenditure for office visits,
- average annual out-of-pocket (OOP) expense for hospital visits,
- average annual out-of-pocket expense for facility visits,
- average annual out-of-pocket expense for office visits, and
- the proportion with large out-of-pocket expenditures.

To determine the sample sizes required for the linked designs, the variance had to be modeled for the unlinked, self-weighting design.

The NMCUES estimation approach is to construct means in terms of total 1980 person-years rather than in terms of all persons ever existing in 1980. For domain k, the mean utilization or expenditure per person-year was estimated as:

$$\bar{y}_k(\text{NMCUES}) = \sum_{i \varepsilon S} W(i)\ \delta_k(i)\ Y(i)\ /\ \sum_{i \varepsilon S} W(i)\ T(i)\ \delta_k(i) \qquad (5\text{-}4)$$

where

$\delta_k(i)$ is 1 if the i-th person belongs to the k-th domain and 0 otherwise,

$Y(i)$ is the response of the i-th person,

$W(i)$ is the analysis weight for the i-th person, and

$T(i)$ is the time-adjustment factor for the i-th person.

The numerator estimates total expenditures or utilization and the denominator the average annual number of persons in the population (i.e., the total person-years). The time adjustment factor, T(i), is the total days in the year that person i is eligible for NMCUES divided by the total days in the year.

Large out-of-pocket expenditures were defined as "annualized" out-of-pocket expenditures of \$200.00 or more. The annualized out-of-pocket expenditure is the annual out-of-pocket expenditure divided by the fraction of 1980 that the person was eligible. For domain k, the proportion with large out of pocket expenditures was estimated as:

$$\bar{y}_k(\text{NMCUES}) = \sum_{i \varepsilon S} W(i)\, T(i)\, \delta_k(i)\, Y(i) \Big/ \sum_{i \varepsilon S} W(i)\, T(i)\, \delta_k(i) \qquad (5\text{-}5)$$

where Y(i) is 1 if the person had large out-of-pocket expenditures and 0 otherwise.

The variance of $\bar{y}_k$(NMCUES) was derived assuming a three-stage household survey design patterned after the 1980 NMCUES sample design, with SMSA/County-sized PSUs and area segments (SEGs) selected as noncompact clusters of dwelling units. Using this approach, the variance of $\bar{y}_k$(NMCUES) may be modeled as:

$$\text{Var}[\bar{y}_k(\text{NMCUES})] = \sigma_k^2(\text{PSU})/r + \sigma_k^2(\text{SEG})/r\bar{s} + \sigma_k^2(\text{HH})/r\bar{s}\bar{t} \qquad (5\text{-}6)$$

where

r is the number of PSUs,

$\bar{s}$ is the average number of segments per PSU,

$\bar{t}$ is the average number of responding households per segment,

$\sigma_k^2(\text{PSU})$ is the between-PSU, within-stratum variance component for domain k,

$\sigma_k^2(\text{SEG})$ is the between-segment, within-PSU variance component for domain k, and

$\sigma_k^2(\text{HH})$ is the between-household, within-segment variance component for domain k.

A variance components estimation program (Shah, 1979) was applied to the 1980 NMCUES data to estimate the variance components for PSUs, segments, and households. Table 5-1 presents the percentage of NMCUES expenditure and utilization variation explained by each component for the various types of service statistics for the total population and Medicaid recipients.

These three-stage variance component estimates were used to estimate the variances that would be achieved by self-weighting NMCUES designs with 6,000 and 10,000 responding households. The terms remaining to be specified in the variance expression presented in equation (5-6) are the number of PSUs (r), the average number of segments sampled per PSU ($\bar{s}$), and the average number of households sampled per segment ($\bar{t}$). For modeling purposes, the RTI general purpose sample was assumed for the next NMCUES, which contains 102 PSUs ($r = 102$). A future NMCUES should experience no worse than the

Table 5-1. Percent of NMCUES Expenditure and Utilization Variation by Domain and Type of Service

Domain	Statistic	Percent of Variation		
		PSU	Segment	Household
Total	Hospital Visits	0.61	0.07	99.32
	Facility Visits	1.34	5.17	93.49
	Office Visits	0.66	2.02	97.32
	Hospital Charges	0.02	0.28	99.70
	Facility Charges	0.59	3.38	96.03
	Office Charges	0.03	3.28	96.69
	Hospital OOP Expenses	0.02	0.65	99.33
	Facility OOP Expenses	0.48	0.92	98.60
	Office OOP Expenses	0.02	6.31	93.67
	Proportion With Large OOP Expenses	0.02	5.93	94.05
Medicaid	Hospital Visits	0.07	0.73	99.20
	Facility Visits	0.41	3.60	95.99
	Office Visits	0.49	0.56	98.95
	Hospital Charges	0.02	0.83	99.15
	Facility Charges	0.03	1.53	98.44
	Office Charges	0.50	0.02	99.48
	Hospital OOP Expenses	0.19	0.03	99.78
	Facility OOP Expenses	0.03	0.03	99.94
	Office OOP Expenses	0.02	0.20	99.78
	Proportion With Large OOP Expenses	0.25	2.06	97.69

nonresponse and attrition encountered by the 1980 NMCUES. Therefore, the 1980 NMCUES experience was ratio adjusted to produce the sample sizes required for the 6,000 and 10,000 household designs. Since the 1980 NMCUES had been designed to be optimal with respect to the number of selections per segment, the number of responding households per segment was set to the value that the 1980 NMCUES achieved or $\bar{t} = 8$. The total number of segments in the 6,000 responding OBRU design would then be 750 ($r\bar{s} = 750$) and 1,250 for the 10,000 responding household design ($r\bar{s} = 1{,}250$).

These estimated variances were used as precision criteria for the other designs that were investigated in this study. Table 5-2 presents the results of this variance modeling activity for the two domains of interest and the ten outcome measures. For convenience, percent relative standard errors are presented rather than variances. The <u>percent relative standard error</u> is the standard error (the square root of the variance) divided by the parameter being estimated, expressed as a percentage. The percent relative standard errors achieved by the 6,000 household design are sufficient for the estimates based upon the total domain but the increased precision that the 10,000 household design achieves for the Medicaid estimates would be desirable.

5.2.4 Cost Modeling

Since cost comparisons would be needed between the unlinked designs and the linked designs, a systematic method was needed to generate the costs for all designs. The approach that was used in this study was to develop unit costs by task for each design. The

Table 5-2. Estimated Means and Percent Relative Standard Errors for the Unlinked NMCUES Design with 6,000 and 10,000 Responding Households

Domain	Statistic	$\bar{y}_k$(NMCUES)	Percent Relative Standard Erro	
			6,000 Households	10,000 Househol
Total	Hospital Visits	0.18	3.11	2.61
	Facility Visits	0.86	4.92	4.25
	Office Visits	4.18	2.02	1.69
	Hospital Charges	362.04	6.22	4.84
	Facility Charges	50.56	4.95	4.11
	Office Charges	117.71	2.42	1.88
	Hospital OOP Expenses	33.10	12.08	9.39
	Facility OOP Expenses	9.77	4.82	3.99
	Office OOP Expenses	53.70	2.43	1.89
	Proportion With Large OOP Expenses	0.24	7.03	5.47
Medicaid	Hospital Visits	0.33	6.63	5.20
	Facility Visits	1.36	7.70	6.27
	Office Visits	5.21	5.59	4.63
	Hospital Charges	691.56	13.56	10.55
	Facility Charges	78.09	7.45	5.80
	Office Charges	139.60	7.27	6.04
	Hospital OOP Expenses	36.18	29.97	23.98
	Facility OOP Expenses	7.39	20.80	16.19
	Office OOP Expenses	23.10	9.57	7.44
	Proportion With Large OOP Expenses	0.11	22.79	18.32

NMCUES tasks that were included in the modeling were the basic sampling, weighting, data collection, and data processing tasks. The unit costs that were developed for each task were fixed costs, PSU level costs, segment level costs, and reporting unit level costs.

The first step in the process was to document what the actual cost experience had been for the 1980 NMCUES. An early decision was made to include only direct costs in the modeling. Indirect costs are the mechanisms whereby contractors recover costs that cannot be directly charged to a project, such as the costs for administration and building maintenance. Since these indirect costs vary across contractors as well as accounting procedures used to recover these costs, only direct costs were modeled. For the 6,000 household design, direct costs would be $4,963,013. For the 10,000 household design, the direct costs would be $7,209,409.

5.3. Optimally Allocated Linked Designs

With knowledge of the characteristics of NHIS respondents, there are possibilities for gains due to stratification and to optimally allocating the sample. To investigate these possibilities, five optimally allocated linked household designs were investigated. The first two designs were optimally allocated self-weighting designs, one with the precision of the 6,000 household unlinked design and the other with the precision of the 10,000 household design. Next, the self-weighting requirement was dropped and two optimally allocated designs were developed, one using the 6,000 household constraints and the second using the 10,000 household constraints. Since the main reason to increase the sample size to 10,000 households was to obtain improved precision for the smaller domains such as Medicaid recipients, the decision was made to investigate one last design in which the precision constraints for the total population were set to those achieved by the 6,000 household design, and the precision constraints for the Medicaid subpopulation were set to those achieved by the 10,000 household design.

5.3.1 The Optimization Problem

When NHIS households are used as the sampling units for the NMCUES, there is much useful information about the households from the NHIS interview. Most of this information is individual-level such as age, race, sex, relationship to head, limitation of activity, bed disability days, perceived health status, medical conditions, education level, marital status, and employment. Since NMCUES samples entire households to facilitate family-level analysis, these data would have to be aggregated to the household level to be used for stratification purposes.

Stratification of the NHIS sample prior to selecting the NMCUES sample could serve the dual purpose of providing control over the distribution of the sample while increasing the precision of survey estimates. The variance of survey estimates is reduced and hence the precision increased through stratified sampling when the strata are formed to maximize the between-stratum variation and minimize the within-stratum variation. Variables to use for stratification are those that result in homogeneity of the units within strata and associated heterogeneity between the stratum means.

Time constraints prevented the examination of 1980 NMCUES data to determine what variables might best be used for the stratification of the NHIS sample prior to NMCUES sample selection. Rather, variables that were generally known to be good predictors of health care utilization and expenditures were used for stratification. Specifically, black/nonblack, aged/nonaged, poor/nonpoor, and self-perceived health status (healthy/non-healthy) variables were used for stratification in this demonstration. Sample size limitations of the 1980 NMCUES data base used to estimate variance components required collapsing the black strata over the poor/nonpoor variable, resulting in eight nonblack strata and four black strata.

To illustrate the advantages of an optimal-allocation approach, five optimal designs were developed. The domains that were included in the optimizations were the total population and Medicaid recipients. For use in stratification, household-level variables were defined that denoted the race (black versus nonblack), poverty status (above or below 150 percent of the value defined as poverty), aged status (containing no person 65 or over versus at least one), and health status (containing no person with poor or fair health versus at least one). The optimization was performed for nine utilization and expenditure means and the subpopulation proportion burdened with large out-of-pocket expenses.

5.3.2 Variance Modeling for the Stratified Design

The first step in developing optimally allocated designs was modeling the variance for a stratified household sample drawn from the first-phase NHIS sample households. To describe the variances for the linked household design, the characteristics of the NHIS sample need to be described, since the NHIS would be used as the NMCUES frame. The redesigned NHIS will have the same target population as the NMCUES. To represent this target population, the NHIS will include 200 sample PSUs and 8,750 segments from these PSUs. The segments will contain 40 addresses on the average with 6 addresses selected for inclusion in the NHIS. The sample segments will be partitioned into 52 sets which will be allocated to weeks so that each weekly sample is a valid national sample. An added feature of the NHIS is that blacks will be oversampled at a rate of 1.4 times that of nonblacks. Finer details of the structure of the redesigned NHIS had not been developed at the time this study was conducted.

Using a stratified sampling approach to subsampling NHIS sample households, NMCUES would estimate the mean for domain k as:

$$\bar{y}_k(\text{NMCUES}) = \sum_{h=1}^{H} \hat{\pi}_k(h)\ \bar{y}_k(h) \tag{5-7}$$

where

$\bar{y}_k(h)$ is the NMCUES estimated mean for stratum h,

$\hat{\pi}_k(h)$ is the NHIS-estimated fraction of the k-th subpopulation total person-years associated with the h-th stratum, and

H is the number of sample strata.

For the nine utilization and expenditure measures, the stratum mean is estimated as

$$\bar{y}_k(h) = \sum_{i\varepsilon h} W(i)\ \delta_k(i)\ Y(i)\ /\ \sum_{i\varepsilon h} W(i)\ \delta_k(i)\ T(i) \tag{5-8}$$

where

$Y(i)$ is the response of the i-th person,

$W(i)$ is the analysis weight of the i-th person,

$\delta_k(i)$ is one if the i-th person belongs to the k-th domain and zero otherwise, and

$T(i)$ is the fraction of the year that the i-th person was eligible for NMCUES.

For the proportion burdened with large out-of-pocket expenses, the stratum mean is estimated as

$$\bar{y}_k(h) = \sum_{i \varepsilon h} W(i)\ \delta_k(i)\ T(i)\ Y(i)\ /\ \sum_{i \varepsilon h} W(i)\ \delta_k(i)\ T(i) \qquad (5\text{-}9)$$

where $Y(i)$ is one if the annualized out-of-pocket expenses are large (> \$200) and zero otherwise.

To simplify modeling the variance, it was assumed that NHIS oversampling of blacks would be at the last stage and that black/nonblack would be a stratification variable. Under this assumption, the variance of the stratified estimate can be modeled as:

$$\begin{aligned} \mathrm{Var}[\bar{y}_k(\mathrm{NMCUES})] &= \mathrm{Var}_{\mathrm{NHIS}}\ [E\ (\bar{y}_k|\mathrm{NHIS})] + E_{\mathrm{NHIS}}\ [\mathrm{Var}\ (\bar{y}_k|\mathrm{NHIS})] \\ &= \mathrm{Var}_{\mathrm{NHIS}}\ [\bar{y}_k(\mathrm{NHIS})] + E_{\mathrm{NHIS}}\ \{\sum_{h=1}^{H} \pi_k^2(h)\ S_k^2(h)[1-f(h)]/m(h)\} \end{aligned}$$

or

$$\begin{aligned} \mathrm{Var}[\bar{y}_k(\mathrm{NMCUES})] &\hat{=} D_w(k)\ [\sigma_k^2(\mathrm{PSU})/r + \sigma_k^2(\mathrm{SEG})/r\bar{s} + \sigma_k^2(\mathrm{HH})/r\bar{s}\bar{t}] \\ &+ \sum_{h=1}^{H} \pi_k^2(h)\ S_k^2(h)\ [1-f(h)]\ /\ E[m(h)] \end{aligned} \qquad (5\text{-}10)$$

where

- $f(h)$ is the NMCUES subsampling rate for stratum h or $m(h)/n(h)$,
- $m(h)$ is the NMCUES stratum h household sample size,
- $n(h)$ is the NHIS stratum h household sample size,
- $D_w(k)$ is the design effect for NHIS unequal weighting for the k-th domain,
- $\sigma_k^2(PSU)$ is the between-NHIS-PSU variance component for domain k,
- $\sigma_k^2(SEG)$ is the between-NHIS-segment, within-NHIS-PSU, variance component for domain k,
- $\sigma_k^2(HH)$ is the between-NHIS-household, within-NHIS-segment, variance component for domain k, and
- $S_k^2(h)$ is the stratum h variance for domain k.

Again, the variance components computed from the 1980 NMCUES were used to estimate the NHIS components. A Taylor Series approximation for the simple random sampling variance of a combined ratio estimator was used to estimate $S_k^2(h)$.

The expected NMCUES sample size from the h-th stratum can be expressed as

$$E[m(h)] = r\bar{s}\bar{t}\ f(h)\ \pi'(h) \qquad (5\text{-}11)$$

where $\pi'(h)$ is the expected fraction of the NHIS sample from the h-th stratum or

$$\pi'(h) = M(h)\ o(h) \Big/ \sum_{h=1}^{H} M(h)\ o(h) \qquad (5\text{-}12)$$

and $M(h)$ is the population count of households in stratum h. The $o(h)$ term represents the NHIS oversampling of black strata; that is, $o(h) = 1.0$ for nonblack strata and $o(h) = 1.4$ for black strata.

Presuming that black/nonblack is used as a stratification variable with equal probability sampling within strata, the design effect for unequal weighting in domain k estimation may be modeled as

$$D_w(k) = \pi_B^2 / \theta_B + \pi_{NB}^2 / \theta_{NB} \tag{5-13}$$

where π_B and π_{NB} are the proportion of blacks and the proportion of nonblacks in the population and θ_B and θ_{NB} are the proportion of blacks and the proportion of nonblacks in the NHIS sample. Note that

$$\theta_B = 1.4\ \pi_B / (1.4\ \pi_B + \pi_{NB}) \tag{5-14}$$

and

$$\theta_{NB} = \pi_{NB} / (1.4\ \pi_B + \pi_{NB}) \tag{5-15}$$

since the NHIS will oversample blacks by 1.4 times the rate at which they occur in the population. Hence, $D_w(k)$ may also be expressed as

$$D_w(k) = 1 + (0.16\ \pi_B\ \pi_{NB} / 1.4) \tag{5-16}$$

For convenience sake, relative variance components were used in the optimization. To model the relative variances, note that

$$RV_k(\text{NMCUES}) = \text{Var}[\bar{y}_k(\text{NMCUES})] / \bar{y}_k^2(\text{NMCUES}) \tag{5-17}$$

For domain k, the relative variance of a mean estimated using the linked household design can be expressed as:

$$RV_k(\text{NMCUES}) = \sum_{\ell=1}^{H} RV_k(\ell) / m(\ell) + \sum_{\ell=H+1}^{H+2} RV_k(\ell) / m(\ell) \tag{5-18}$$

where $\ell = 1,2,\ldots,H$ are the second-phase strata used in selecting the NMCUES subsample and H+1 and H+2 are the first-phase segment and PSU sampling stages.

5.3.3 Cost Modeling

The next step in developing optimally allocated designs was modeling the cost components associated with each second-phase NMCUES stratum and each stage of the first-phase NHIS design. Let $C(\ell)$ represent the variable unit cost for a selection from level ℓ. Then the optimization problem may be stated as follows:

$$\text{Minimize CV(NMCUES)} = \sum_{\ell=1}^{H+2} m(\ell)\, C(\ell) \tag{5-19}$$

subject to

$$\sum_{\ell=1}^{H+2} RV_k(\ell) / m(\ell) \le RV_k^* \text{ for } k=1,2,\ldots,K;$$

$$m(\ell) \ge 0 \text{ for } \ell=1,2,\ldots,H+2;$$

$$200 \le m(H+2) \le m(H+1); \text{ and}$$

$$m(\ell) \le m(H+1) \text{ for } \ell=1,2,\ldots,H.$$

CV(NMCUES) is the total variable cost for NMCUES and RV_k^* is the relative variance constraint established for the k-th domain.

The variable costs for the PSU stage of sampling [C(H+2)] and the segment level of sampling [C(H+1)] were obtained by modifying the task-level unit costs produced as a part of the cost modeling of the unlinked household design. The unit costs for the subsampled households within NHIS-defined strata vary depending upon the response rate and movement rates within the strata. The 1980 NMCUES experience was used to estimate the roundwise rates at which ineligibles, nonrespondents, and movers would be encountered and to develop the household-level cost component for each of the 12 strata. Unit costs were developed for movers, tracing, interviewing ineligibles, and interviewing outside and inside the clusters and used in forming the overall unit costs for each stratum.

5.3.4 Self-Weighting Optimally Allocated Designs

The first type of design that was investigated was a stratified, self-weighting, linked household design. Using this design, the variance would be expressed as in equation (5-10) where $f(h) = f/o(h)$. The factor f is the overall subsampling rate desired for the NMCUES subsample of the NHIS after NHIS oversampling is removed. The Chromy optimization procedure was used to obtain optimum values for the number of PSUs, the average number of segments to sample per PSU, and the NMCUES subsampling rate to be used within the sample segments (r, $\bar{s}$, and f).

The optimization was performed twice. When the variance constraints associated with the 6,000 household unlinked design were used, the optimal solution was 102 PSUs; 1,258 segments; and 5,980 responding households. With a subsampling rate f of 83 percent, black strata would be subsampled at a 59 percent rate ($f/1.4$) and nonblack strata at the 83 percent rate. When this design is used, the total cost for the design is $4,844,013 as compared to $4,963,013 for the unlinked design with the same precision.

When the variance constraints associated with the 10,000 household unlinked design were used, the optimal stratified linked household design had 103 PSUs; 2,117 segments; 9,960 responding households; and a subsampling rate f of 82 percent. Allowing for the NHIS oversampling implies that black strata would be subsampled at a 58 percent rate and nonblack strata at the 82 percent rate. When this design is used, the total cost is $6,931,233 as compared to $7,209,409 for the unlinked design with equivalent precision.

5.3.5 Nonself-Weighting Optimally Allocated Designs

The next set of designs that was investigated was the stratified linked household designs without the self-weighting requirement. The advantage of this type of design is that heavy utilizers of health care services can be identified and oversampled. Since optimization occurs over PSUs (r), segments ($r\bar{s}$), and NMCUES strata (h=1,2,...,H), the stratified linked sample has H+2 design levels.

The first design that was investigated was an optimally allocated design with the precision constraints of the unlinked 6,000 household design for the total and Medicaid domains. The optimal solution used 98 PSUs; 1,152 segments; and 5,880 responding households with subsampling rates ranging from 57 to 100 percent. In general, the unhealthy and nonblacks were sampled at a higher rate. Greater percentages of the NHIS sample of nonblacks were selected than blacks as the result of the fact that blacks occur in the NHIS sample at a rate 1.4 greater than nonblacks. The total cost for this design is $4,770,353 as compared to $4,963,013 for the unlinked 6,000 household design and $4,844,013 for the equivalent self-weighting optimally allocated design.

The next design that was investigated was an optimally allocated design with the precision of the 10,000 household unlinked design for the total and Medicaid domains. The optimal solution used 106 PSUs; 1,811 segments; and 9,717 responding households with subsampling rates ranging from 59 to 100 percent. The total cost for the design was $6,758,063 as compared to $7,209,409 for the 10,000 household unlinked design and $6,931,233 for the equivalent self-weighting optimally allocated design.

For household samples drawn from area frames, there is little information available for use in sample stratification, and what information is available is for geographical areas rather than households or dwelling units. To obtain the required sample sizes for small domains, a larger than otherwise needed sample size is frequently used. With household-level stratification information, these small domains can be oversampled without having to correspondingly increase the size of the total sample.

To illustrate this advantage for the NMCUES, an optimally allocated design was created where the precision of the 10,000 household design was specified for the Medicaid domain but the precision of the 6,000 household design was deemed satisfactory for total population estimates. These constraints result in an optimal design with 95 PSUs; 2,092 segments; and 7,228 responding households with subsampling rates ranging from 32 to 100 percent. The total cost for the (6,000/10,000) design was $5,601,533 which compares quite favorably with the $6,758,063 costs for the comparable non-self-weighting design with 10,000 household constraints for both the total and Medicaid domain statistics.

5.4. Comparison of the Linked and Unlinked Designs

Three types of sample designs have been described in this chapter, including two unlinked designs, two optimally allocated self-weighting linked household designs, and three optimally allocated nonself-weighting linked household designs. Table 5-3 summarizes the sample sizes and costs for the designs investigated for potential use in future NMCUES. The optimally allocated designs contrast quite favorably with the unlinked designs in terms of cost.

Table 5-3 also gives the months that the NHIS sample would have to be aggregated to obtain the required number of sample segments from the specified number of PSUs. These estimates of aggregation time were based upon the assumptions that NHIS would include 8,750 segments and 200 PSUs for an average of 43.75 segments per PSU in a year's time, and that the NMCUES would be selected from the 90 percent that were conducted by personal interview. The aggregation times range from 1.5 to 6.7 months with the longer periods of aggregation found for the optimally allocated designs. The modeling of movement was only approximate so that the costs associated with movement may be understated, particularly for designs that aggregate over a longer period of time. More attention could be given to cost modeling of movement as the time between the NHIS and NMCUES increases.

Linkage of the NMCUES to the NHIS has the unique advantage of knowing the names, addresses, and personal characteristics of sample households in advance of data collection. The design with the most potential for exploiting this knowledge is the stratified nonself-

Table 5-3. Sample Size Summary for the Alternate NMCUES Designs

Design Type	Sample Sizes			Aggregation Time	Direct Costs
	PSUs	Segments	Households		
Unlinked Designs:					
6,000 Households	102	750	6,000	N/A	4,963,013
10,000 Households	102	1,250	10,000	N/A	7,209,409
Linked Optimally Allocated Designs:					
S.W. 6,000/6,000	102	1,258	5,980	3.8	4,844,013
S.W. 10,000/10,000	103	2,117	9,960	6.3	6,931,233
N.S.W. 6,000/6,000	98	1,152	5,880	3.6	4,770,353
N.S.W. 10,000/10,000	106	1,811	9,717	5.2	6,758,063
N.S.W. 6,000/10,000	95	2,092	7,228	6.7	5,601,533

weighting optimally allocated design. Research could be conducted to produce such a design for the next NMCUES. This research would determine the domains and statistics of interest to the survey and the most appropriate set to include in the optimization. The gain from the use of an optimally allocated design should far exceed the costs of developing such a design.

5.5 Concluding Remarks

The NMCUES has many small analysis domains of interest including the Medicaid population, the Medicare population, the aged, the poor, and blacks. The need for these domain analyses has led NMCUES in the past to use large self-weighting samples to obtain adequate precision for these small domains. This approach resulted in greater precision than was needed for large domains such as the nonaged or white domains. Without linkage, however, this is the best approach possible since household characteristics are not available for use in sampling.

With linkage to the NHIS, there is a plethora of information about the households that can be used to create an optimally allocated design with increased precision for selected domains. This design strategy could be pursued much further than was possible in this demonstration. Precision constraints could be set for a larger group of policy-relevant domains. The stability of the variance components would also need to be considered and the accuracy of the cost components. Finally, the effect of the length of aggregation of the NHIS sample could be built into the cost modeling.

By relaxing the self-weighting condition, an optimally allocated design can be created that obtains the desired precision for a small domain by oversampling from strata where domain members are concentrated. To repeat the example cited earlier, if the required variance constraints for the Medicaid domain are those achieved by the 10,000 household unlinked design, then the self-weighting stratified linked design that would be used is the one that achieves the variance constraints of the 10,000 household unlinked design for all domains. If the variance constraints achieved by the 6,000 household unlinked design were acceptable for the total population, the nonself-weighting stratified optimally allocated linked design can achieve both sets of variance constraints by oversampling strata with a high concentration of Medicaid recipients. The survey costs with the nonself-weighting approach would be $5,601,533 as compared to $6,931,233 with the self-weighting design.

In constructing an optimal design, careful attention needs to be given to the reporting domains to be included in the optimization. The survey planner is assured of acceptable levels of precision for those statistics and domains that are included in the optimization. The precision for other statistics and other domains will depend upon the extent to which they are related to the statistics and domains included in the optimization.

The disadvantage of the optimally allocated nonself-weighting approach is associated with estimation for domains and/or statistics not included in the optimization. The nonself-weighting 6,000/10,000 design produces estimates of the desired precision for the total utilization and total expenditures statistics by oversampling from the unhealthy strata. If total income is being estimated instead, this design may or may not yield estimates of the desired precision since the design did not control for the precision of income estimates. Alternatively, if total utilization or total expenditures are being estimated for a domain not included in the optimization, such as the college-educated domain for instance, then again the design may or may not yield estimates of the desired precision. The precision of estimates for domains and/or statistics not included in the optimization will depend upon the extent to which the statistics and/or domains are related to the statistics and domains included in the optimization.

In practice, most surveys have multiple domains that are of interest and a diversity of statistics to report. This does not imply that a nonself-weighting optimally allocated design is unacceptable. In this situation, several estimates can be considered simultaneously where the estimates are chosen by classifying their variance properties and selecting a typical variance model from each class. Similarly, the domains to include in the optimization can be chosen by listing the important domains of interest and selecting domains that represent diverse groups of the population. Care should be taken that the extremes of the domains of interest are represented in the set of domains subject to optimization, since the extreme groups are usually the rarest and hence an adequate sample size will not be obtained unless special steps are taken. Thus, a survey particularly interested in contrasting health expenditures for different income groups would want to include the poor and the wealthy as domains in the optimization. With a large proportion of the population middle income, there may be no need to explicitly include them as a domain, particularly if the total population is included as a domain in the optimization.

References

Cochran, W. G. (1977). Sampling Techniques, New York: John Wiley and Sons.

Cox, B. G., R. E. Folsom, T. G. Virag, and W. F. Refior (1983). Design Alternatives for Integrating the NMCUES with the NHIS, Series 2 Report, National Center for Health Statistics, Hyattsville, Maryland (in press).

Folsom, R. E. Jr., R. L. Williams, and J. R. Chromy (1980). Optimum Design of a Medical Care Expenditure and Utilization Survey Involving a Provider Record Check, Report No. RTI/1725/01-06S, Contract No. 233-78-2102, National Center for Health Statistics, Hyattsville, Maryland.

Hansen, M. H., W. N. Hurwitz, and W. G. Madow (1953). Sample Survey Methods and Theory; Volume I Methods and Applications, New York: John Wiley and Sons.

Shah, B. V. (1979). VMCPNLS: Program to Compute Variance Components, Research Triangle Institute, Research Triangle Park, North Carolina.

6

A Comparison of Household and Provider Reports of Medical Conditions

A component of the analysis plan for many health care surveys is the classification of individuals in terms of their health status and the disease or disability conditions to which they are subject. This information is usually obtained by direct interview methods in which each household member is asked to provide information. To the extent that individuals do not recall or do not accurately understand the conditions to which they are subject, the household survey data will be in error. Other more accurate sources of condition data are available, however. Medical care practitioners and providers document their professional appraisal of the patient's medical conditions in their records. With direct access to their records, they need not rely on memory and hence, can provide more reliable data about the health status and medical conditions of their patients. With 87 percent of the general population insured (National Center for Health Statistics, 1983), public and private health care insurers are another important source of information about an individual's health, particularly expenditures and utilization, although condition data may be available as well.

These alternative data sources cannot replace the household interview entirely since medical conditions can exist that result in work loss or days in bed but for which no provider is seen. In addition, the high cost of obtaining data from health care providers and insurers often precludes the use of these sources of information to supplement the household report. Thus, most health care surveys must rely partially or totally upon data collected from sample individuals.

Under these circumstances, it becomes important to understand the nature and accuracy of the data that can be obtained in household surveys. The lack of this sort of information prompted the use of record check components to supplement the household survey data obtained in the National Medical Care Expenditure Survey (NMCES). The Medical Provider Survey (MPS) component of NMCES obtained ambulatory visit data and hospital stay data from the medical providers of sample individuals. Eligible for inclusion in the MPS were those medical providers or usual sources of care of sample individuals that were medical doctors (MDs), doctors of osteopathy (DOs), and hospitals and other facilities run by MDs or DOs. The visit and stay data obtained in the MPS were matched to the visits and stays reported in the household interview. This chapter presents the results of an investigation of the extent to which the household and provider reports agree with respect to the medical conditions associated with health care visits (Cox and McGrath, 1981 and Cox and Gridley, 1981).

Because the Medical Provider Survey was designed to over-represent individuals who were thought to be poor reporters, weighted estimates are presented that reflect the MPS probability sampling. The data used in the investigation were the condition data for those household reported visits and stays that had been matched to provider reported visits. To match the two sets of data, individual treated, provider visited, visit or stay date, and charge were used, but not the condition associated with the visit (Cooley and Cox, 1981).

For each visit reported in the household survey, the interviewer was instructed to ask the respondent,

> For what condition did (PERSON) [see or talk to (PROVIDER)/have (SERVICE)/go to (PLACE)] on (DATE)? Any other condition?

Thus, the respondent was asked to report the <u>conditions</u> that led to his making the medical visit. A similar approach was used when the respondent reported a hospital stay.

For each visit reported in the Medical Provider Survey for a sample individual, the provider was asked,

> What diagnoses/complaints were treated during this visit? If no diagnosis was made, please list conditions or symptoms treated.

Hence, the provider was instructed to record his <u>diagnoses</u> rather than the reasons the respondent gave for making the visit. In general, one would expect that the condition that led the respondent to seek treatment would be related to the diagnoses made by the provider. This study explores the relationship between the two sets of data.

6.1 Recoding of the Household and Provider Reported Condition Data

As reported by the household, condition data are subject to response errors caused by respondent inability to describe the exact conditions associated with an ambulatory visit or hospital stay or inaccurately reporting the conditions. In many cases, the respondents can provide symptom data only; for example, the respondent may say he visited a provider because he had "heart trouble" or "difficulty breathing." For the matched set of NMCES household and MPS provider visits, the condition data reported by the household can be compared with the diagnoses reported by their medical providers. Since there are thousands of conditions that could be reported, the investigation of the relationship between the two reports was done after the data had been recoded to produce a smaller number of condition categories. The recoding operation was complicated by the fact that two different coding systems were used for the two sets of NMCES data: the household data were coded using the National Health Interview Survey (NHIS) system and the provider data using the American adaptation of the International Classification of Diseases (ICDA). The NHIS coding system is similar but not identical to the ICDA system.

The objective of the recoding operation was to define categories that were internally consistent and as diverse from one another as possible, in addition to being of reasonable size and of general analytical interest. The basic idea was that individuals might not have an accurate knowledge of exact conditions but would likely know the biological system affected and the condition type.

Analytical interests caused some related conditions to be partitioned into different recodes; hence, some recode categories are related in the sense that they affect the same biological system or are similar in nature. The final consideration guiding the development of recodes was the size of the resultant categories. Since these categories were to be used in statistical analyses, a minimum size, in terms of matched visits for each recode classification, was required in order to be able to model the relationship between the household-reported data and the provider-reported data.

The National Health Interview Survey (NHIS) developed a series of condition recodes in 1969 for recoding the household data obtained in the NHIS interview. Since the NHIS and NMCES instruments are similar and the NHIS coding system was used for classifying the condition data obtained in the NMCES household interviews, it seemed appropriate to adapt the NHIS recodes for use in NMCES. The recode categories were required to be disjoint and complete in the sense that each NHIS condition code would fall into one, and only one, recode category. Having defined recode categories in terms of NHIS condition codes, the NHIS coding manual (USDHEW, 1969) was used in conjunction with the ICDA coding manual (USDHEW, 1968) to determine the equivalent ICDA codes for each category. Descriptions of the 62 codes are provided in Table 6-1.*

---

* Originally 64 categories were defined; two categories were combined with others at a later stage. The categories were not renumbered so that the recodes are numbered from 1 to 64 with two numbers, 41 and 47, not used.

As far as possible, the recode categories were defined so that when the household respondent gave a condition that was equivalent to the provider's diagnosis for a visit, the visit would be classified into the same recode category for the two reports. Because there is not a one to one correspondence between the two coding systems for all condition codes, some discrepancies between the NHIS and ICDA definitions for the categories were unavoidable. This was principally a problem in defining the recode categories related to musculoskeletal conditions, injuries, and impairments.

It is important to realize that the condition that a household or provider reported could be "no condition." For instance, a visit for a medical checkup would usually result in no conditions being reported. No attempt was made to distinguish between visits for which no condition was reported because there was no condition associated with the visit and those visits for which no condition was given because the household did not know the condition. In both of these situations, the visit was classified as no condition reported.

Both households and providers frequently reported multiple complaints for a medical care visit or stay. In recoding these data, zero-one indicator variables were created for each NMCES recode category in order to preserve the many combinations of diagnoses and conditions that the sample individual and his provider might report. Each visit has a "1" for the indicator variable associated with a category when the data contained a condition or diagnosis code that fell into the recode category. Separate sets of

Table 6-1. NMCES Condition Recodes

Recode	Description
1	Measles, Rubella, Whooping Cough, Chicken Pox, Mumps
2	Viral Infections, Unspecified
3	Intestinal Infectious Diseases
4	Other Infective and Parasitic Diseases
5	Malignant Neoplasms
6	Benign and Unspecified Neoplasms
7	Diseases and Conditions of the Endocrine Glands (Excluding Diabetes)
8	Diabetes (Mellitus)
9	Nutritional and Metabolic Diseases and Conditions
10	Anemias
11	Other Diseases of Blood and Blood Forming Organs
12	Specified Mental Disorders
13	Neuroses
14	Alcoholism
15	Nervousness, Depression, Special Symptoms

16	Headache and Migraine
17	Diseases and Conditions of the Central Nervous System
18	Cataract
19	Diseases and Conditions of the Eye (Other Than Cataracts and Viral Diseases of the Conjunctiva)
20	Otitis Media
21	Diseases of the Ear (Other Than Otitis Media), Hearing Impairments
22	Hypertensive Disease
23	Other Diseases and Conditions of the Heart
24	Cerebrovascular Disease
25	Diseases and Conditions of the Circulatory System Excluding the Heart
26	Common Cold
27	Influenza
28	Pneumonia
29	Bronchitis
30	Emphysema
31	Asthma (With or Without Hay Fever)
32	Hay Fever, Without Asthma

Table 6-1. (continued)

Recode	Description
33	Sinusitis
34	Tonsillitis
35	Other Diseases and Conditions of the Respiratory System
36	Ulcer of Stomach and Duodenum
37	Functional and Symptomatic Conditions of Stomach
38	Hernia of Abdominal Cavity
39	Diseases of the Liver, Gallbladder, and Pancreas (Other than Alcoholic Cirrhosis of Liver)
40	Other Diseases and Conditions of the Digestive System
42	Diseases and Conditions of the Urinary System
43	Disorders of Menstruation
44	Menopausal Symptoms, Except Psychosis
45	Diseases and Conditions of the Reproductive Organs
46	Pregnancy Care, Deliveries, and Complications of Pregnancy (Excluding Abortions)

48	Diseases and Conditions of the Skin and Subcutaneous Tissue
49	Arthritis, Lumbago Fibrositis, Mysositis, Rheumatism
50	Displacement of Intervertebral Disc
51	Vertebrogenic Pain Syndrome
52	Other Musculoskeletal Diseases and Conditions (Excluding Impairments)
53	Fractures and Dislocations
54	Sprains and Strains
55	Open Wounds and Lacerations
56	Contusions and Superficial Injuries
57	Complications of Medical and Surgical Procedures
58	Other Injuries
59	Paralysis and Congenital Absences Including Cerebral Palsy
60	Impairments (Except Paralysis and Congenital Absences) Back and Spine
61	Impairments (Except Paralysis and Congenital Absences) Upper Extremities and Shoulders
62	Impairments (Except Paralysis and Congenital Absences) Lower Extremities and Hips With Any Other Site
63	All Other Impairments
64	Other Diseases and Conditions

62 recode category indicator variables were created for each visit to record the household condition data and provider diagnostic data. Thus, the full information present in the condition and diagnostic reports was preserved instead of attempting to determine which recode category was more important or which constituted the "principal" complaint.

6.2 Relationship Between Household-Reported Conditions and Provider Diagnoses

The fact that multiple complaints could be reported for a medical care visit or stay complicated the statistical analysis of the relationship between condition as reported by the household and the provider diagnosis. One step in the analysis of the data was determining the frequency with which households and providers reported multiple complaints for a visit. Table 6-2 presents a cross-tabulation of the number of household-reported conditions versus the number of provider-reported diagnoses for the 34,914 matched household and provider reported visits or stays. Note that 0.59 percent of the visits or stays had condition data missing for the household. Missing data for a visit resulted from the inability to link the summary version of a visit to the questionnaire data record where condition data were found.* Of the remaining visits

*The provider records were matched to the summary visit record which contained utilization and expenditure data abstracted from each round's questionnaire and updated by the respondent in subsequent data collection rounds. This matched file formed the basis for the study. When a summary visit record could not be associated with a questionnaire data record, the nonabstracted questionnaire data (e.g., conditions associated with the visit) were lost.

with condition data from the core questionnaire, 92 percent had zero or one conditions reported by the household. Provider diagnosis data were available for all matched visits. On the provider side, 77 percent of the visits had zero or one diagnoses reported. An examination of this table reveals that households tended to report fewer conditions than providers. This is easiest seen by noting that the weighted percentage of visits where the provider reported more diagnoses than the household is five times the weighted percentage of visits where the household reports more conditions. Further, only 53 percent of the visits had the household and provider agreeing on the number of complaints.

The second step in the analysis was to examine the frequency with which providers concurred with a condition reported by the household. For each condition recode category, Table 6-3 gives the number of visits for which the household reported the condition recode and the weighted percent of these visits in which the provider reported the same category. For completeness sake, Table 6-3 also gives the number of visits for which the provider reported the diagnosis and the weighted percent of the visits for which the household reported the same category. This table should be examined carefully as it suggests that the relationship between household reported conditions and provider reports of diagnosis may not be as strong as reseachers would prefer.

Table 6-3 also confirms the remark made earlier that households tend to underreport conditions that the physician observes. This is seen most clearly by examining the agreement when "no condition" was

Table 6-2. The Matched Visits Arrayed by the Number of Household-Reported Conditions Versus the Number of Provider-Reported Diagnoses

Number of Provider-Reported Diagnoses	Number of Household-Reported Conditions								
	0	1	2	3	4	5	6	Missing	Total
0	1,284 4.61 53.48 17.90	1,338 3.57 41.41 5.42	140 0.36 4.15 4.90	8 0.01 0.14 2.44	1 0.00 0.01 1.17	0 0.00 0.00 0.00	0 0.00 0.00 0.00	17 0.07 0.81 11.77	2,788 8.61
1	5,496 17.16 25.18 66.70	15,861 46.72 68.57 70.99	1,423 3.66 5.37 50.15	92 0.21 0.31 42.79	14 0.02 0.03 28.15	2 0.00 0.01 64.60	1 0.00 0.00 24.14	142 0.37 0.54 61.87	23,031 68.14
2	993 2.78 16.72 10.80	4,227 11.53 69.41 17.52	814 2.01 12.10 27.58	72 0.17 1.01 33.91	17 0.03 0.15 35.44	0 0.00 0.00 0.00	1 0.00 0.00 24.14	37 0.10 0.60 16.89	6,161 16.62
3	332 0.82 18.64 3.20	1,119 2.70 61.12 4.10	353 0.78 17.78 10.75	34 0.08 1.78 15.90	6 0.01 0.21 13.25	2 0.00 0.04 25.28	1 0.00 0.02 51.72	8 0.02 0.41 3.04	1,855 4.41

4	100	437	146	8	6	0	0	13	710
	0.23	0.93	0.32	0.01	0.01	0.00	0.00	0.03	1.52
	15.05	60.72	21.12	0.61	0.67	0.00	0.00	1.84	
	0.89	1.41	4.41	1.88	14.18	0.00	0.00	4.73	
5	33	133	60	2	2	0	0	2	232
	.09	.25	.11	0.00	0.00	0.00	0.00	0.01	0.46
	18.85	54.80	24.23	0.39	0.62	0.00	0.00	1.11	
	0.34	0.38	1.52	0.36	4.00	0.00	0.00	0.86	
6	6	44	17	2	1	1	0	0	71
	0.01	0.07	0.03	0.01	0.00	0.00	0.00	0.00	0.12
	11.07	57.54	22.17	7.83	0.78	0.61	0.00	0.00	
	0.05	0.11	0.38	1.97	1.35	10.12	0.00	0.00	
7	12	20	12	1	1	0.00	0.00	1	47
	0.03	0.05	0.02	0.00	0.00	0.00	0.00	0.01	0.10
	25.86	49.17	16.09	2.17	1.73	0.00	0.00	4.97	
	0.10	0.08	0.22	0.45	2.46	0.00	0.00	0.85	
8	6	6	6	1	0	0	0	0	19
	0.01	0.01	0.01	0.00	0.00	0.00	0.00	0.00	0.02
	30.20	30.62	30.94	8.24	0.00	0.00	0.00	0.00	
	0.02	0.01	0.07	0.29	0.00	0.00	0.00	0.00	
Total	8,262	23,185	2,971	220	48	5	3	220	34,914
	25.72	65.82	7.29	0.49	0.07	0.01	0.00	0.59	

The entries are the frequency counts and then the weighted percent, row percent, and column percent.

Table 6-3. Concurrence Between Household and Provider Reports for Each Condition Recode

Recode Number	Description	Number of Visits With the Household Reporting the Category
0	No Condition	8,262
1	Measles, Rubella, Whooping Cough, Chicken Pox, and Mumps	43
2	Viral Infection	206
3	Intestinal Infectious Diseases	156
4	Other Infective and Parasitic Diseases	763
5	Malignant Neoplasms	900
6	Benign and Unspecified Neoplasms	571
7	Endocrine Gland Disorders	168
8	Diabetes	1,315
9	Nutritional and Metabolic Disorders	315
10	Anemias	136
11	Diseases of Blood & Blood Forming Organs	125
12	Specified Mental Disorders	336
13	Neuroses	354
14	Alcoholism	55
15	Nervousness, Depression, Special Symptoms	638
16	Headache and Migraine	184
17	Central Nervous System Disorders	352
18	Cataract	426
19	Other Conditions of the Eye	789
20	Otitus Media	473
21	Other Conditions of the Ear	391
22	Hypertensive Disease	1,819
23	Other Diseases & Conditions of the Heart	1,458
24	Cerebrovascular Disease	154
25	Other Circulatory Conditions	757
26	Common Cold	646
27	Influenza	611
28	Pneumonia	243
29	Bronchitis	330
30	Emphysema	133
31	Asthma	445

Weighted Percent of These Visits With Provider Also Reporting the Category	Number of Visits With the Provider Reporting the Category	Weighted Percent of These Visits With Household Also Reporting the Category
18	2,788	53
51	45	43
7	93	17
37	380	18
35	762	40
58	986	51
27	617	21
47	340	28
74	1,926	50
56	1,037	20
40	365	18
4	72	5
13	494	10
35	720	14
37	108	17
19	781	15
48	380	23
38	732	20
76	558	51
70	1,382	39
56	619	41
40	517	31
63	2,694	43
58	2,388	35
19	96	24
46	1,165	31
10	241	26
8	146	44
45	255	46
50	815	23
63	366	22
34	345	52

Table 6-3. (Continued)

Recode Number	Description	Number of Visits With the Household Reporting the Category
32	Hay Fever	257
33	Sinusitis	262
34	Tonsillitis	216
35	Other Respiratory Conditions	1,170
36	Ulcer of Stomach & Duodenum	212
37	Other Stomach Conditions	121
38	Hernia	254
39	Liver, Gallbladder, & Pancreas Diseases	214
40	Other Conditions of the Digestive System	595
42	Urinary System Conditions	886
43	Disorders of Menstruation	150
44	Menopausal Symptoms	66
45	Conditions of the Reproductive Organs	644
46	Pregnancy Care and Deliveries	791
48	Conditions of the Skin & Subcutaneous Tissue	1,351
49	Arthritis, Lumbago Fibrositis, Mysositis, Rheumatism	702
50	Displacement of the Intervertebral Disc	119
51	Vertebrogenic Pain Syndrome	122
52	Other Musculoskeletal Conditions	626
53	Fractures and Dislocations	971
54	Sprains and Strains	441
55	Open Wounds and Lacerations	620
56	Contusions and Superficial Injuries	422
57	Complications of Medical & Surgical Procedures	169
58	Other Injuries	808
59	Paralysis and Congenital Absences	79
60	Impairments Back and Spine	220
61	Impairments Upper Extremities & Shoulders	139
62	Impairments Lower Extremities & Hips	342
63	All Other Impairments	384
64	All Other Diseases and Conditions	1,377

Weighted Percent of These Visits With Provider Also Reporting the Category	Number of Visits With the Provider Reporting the Category	Weighted Percent of These Visits With Household Also Reporting the Category
32	387	25
28	244	33
53	340	34
49	2,600	22
46	211	44
9	314	4
64	401	42
55	399	29
43	1,255	20
64	1,203	50
51	323	21
18	224	6
52	1,429	23
86	1,747	38
69	2,150	44
52	1,354	27
46	146	42
20	482	6
49	1,256	25
59	797	68
47	623	31
63	567	63
31	520	26
8	128	13
33	709	38
72	96	65
10	86	24
1	21	9
17	165	32
3	100	10
13	4,123	4

reported. Households reported nearly three times more visits as being for no condition than the providers reported. Only 18 percent of the visits that the household reported as "no condition" had the provider also reporting "no diagnosis." The lack of relationship between the two reports is also highlighted by the fact that for those visits for which the provider gave no diagnosis, only 53 percent of the households also reported no condition.

Except for eleven condition recodes, the number of visits that the household reported for a condition category is approximately equal or (as is generally the case) less than the number of visits that the provider reported for the category. Five of these recodes where the household reported significantly more visits than the provider are for viral infections, blood-related conditions, cerebrovascular conditions, common colds, and influenza. It would seem that the general public uses words such as "virus," "cold," "flu," "stroke," and "blood trouble" as a catchall to describe conditions of a similar nature. Another four recodes where the household reported significantly more visits than the provider are associated with the recodes for impairments. The impairment categories were based upon the special "X" codes used by the NHIS. More household reports for the impairment codes may be an artifact of the difference between the NHIS and ICDA coding systems. As discussed in the previous section, there were situations in which the ICDA codes for a category could not be defined in such a manner as to be exactly comparable to the NHIS codes for the category. This was especially true for the impairment recodes.

Over all recode categories, the level of agreement between households and providers is relatively low. The percent of conditions reported by households for which the provider reported a similar diagnosis ranged from 1 to 86 percent with an average value over all conditions of 40 percent. The reverse situation of individuals reporting the same category as their provider had the percent agreement ranging from 5 to 68 percent with an average value over all diagnoses of 30 percent. One would expect households to have difficulty reporting unfamiliar diseases; however, these results show that the level of agreement for diseases that are both common and familiar was lower than might be expected.

Some examples will suffice to illustrate this point. Visits that the household reported as being for measles, rubella, whooping cough, chicken pox, or mumps had the provider agreeing on the complaint 51 percent of the time. Visits for which the household report was anemia had the provider agreeing 40 percent of the time. For household reports of headache or migraine, provider agreement was found for 48 percent. For visits with household reports of asthma, hay fever, and sinusitis, the level of provider agreement was 34, 32, and 28 percent, respectively.

Two categories that one would expect to be reported well would be pregnancy care and diabetes. The level of agreement of the provider when the household reported these two recodes was relatively high, 86 and 74 percent respectively. However, underreporting of visits for these two categories by the household was fairly large. Households reported only 45 percent as many visits for pregnancy care as the provider did and only 68 percent as many visits for diabetes. Part of the underreporting may result from households reporting that a visit was for a checkup when the provider reported the underlying condition requiring the checkup. This is confirmed by the fact that households reported nearly three times more visits with no condition than the providers. Finally, sensitivity of the condition may have produced some underreporting. The recode categories for mental disorders, alcoholism, and menopausal symptoms exhibit low levels of household agreement with the provider report - 6 to 17 percent of the visits have the households reporting the condition when the provider reports the condition - and the provider reported from 1.22 to 3.39 times as many visits for these conditions as the households.

One reason that the level of agreement between provider and household reports is low could be that households cannot report conditions accurately enough so that the visit can be correctly classified with respect to the condition recodes. For instance, if an individual said a visit occurred because he was "having trouble with his vision," the visit would be classified into recode category 19 (Other Diseases and Conditions of the Eye). If the provider said that the visit had a diagnosis of cataract, the visit would be classified into recode category 18 (Cataract). The lack of agreement between the hypothetical reports is caused by the failure of the household to give enough information about the condition to allow it to be classified accurately.

To see if the lack of agreement between the two reports was due to the condition recodes requiring too detailed information from the household, the condition recodes were collapsed into 16 general categories that were essentially equivalent to types of conditions and to major systems of the body. Grouping used by the ICDA were the basis for the collapsed categories, which are defined in Table 6-4. A comparison of the visits by the number of general diagnostic categories reported by the household versus the number reported by the provider demonstrated the same pattern of household underreporting as that observed earlier (see Table 6-5). The household and provider agreed on the number of conditions 57 percent of the time; 35 percent of the time the provider reported more condition categories than the household.

Using the collapsed versions of the condition recodes, cross tabulations were again made of household versus provider data. The results of this analysis are summarized in Table 6-6. Overall, the level of agreement between household and provider condition reports is increased substantially when the collapsed categories are used. The percent of household-reported condition reports for which the provider concurred has an average value of 52 percent as compared with the average value of 40 percent for the recodes with no collapsing. The percent of provider-reported diagnoses for which the household concurred has an average value of 39 percent as compared to 30 percent for the recodes with no collapsing. The lower level of agreement of household with provider reports is again due to the substantial underreporting of conditions by the households.

The improvement in the level of agreement between the two reports is dramatic but there is still substantial disagreement remaining. This disagreement is difficult to explain because of the

Table 6-4. Generalized Version of the Condition Recodes

General Condition Category	Title	Condition Recodes Contained in the Category
1	Infective and Parasitic Diseases	1-4
2	Neoplasms	5,6
3	Endocrine, Nutritional, and Metabolic Diseases and Conditions	7-9
4	Diseases of the Blood and Blood-Forming Organs	10,11
5	Mental Disorders	12-16
6	Diseases and Conditions of the Nervous System and Sense Organs	17-21
7	Circulatory Diseases and Conditions	22-25

8	Respiratory Diseases and Conditions	26-35
9	Digestive Diseases and Conditions	36-40
10	Genitourinary Diseases and Conditions	42-45
11	Pregnancy Care and Complications of Pregnancy, Childbirth, and the Puerperium	46
12	Diseases and Conditions of the Skin and Subcutaneous Tissue	48
13	Diseases and Conditions of the Musculoskeletal System and Connective Tissue	49-52
14	Accidents, Poisonings, and Violence	53-58
15	Paralysis, Congenital Absences, and Impairments	59-63
16	Other Diseases and Conditions	64

Table 6-5. The Matched Visits Arrayed by the Number of Household-Reported Conditions Versus the Number of Provider-Reported Diagnoses for the General Condition Categories

Number of Provider-Reported Diagnoses	Number of Household-Reported Conditions							
	0	1	2	3	4	5	Missing	Total
0	1,284	1,361	122	3	1	0	17	2,788
	4.61	3.63	0.30	0.00	0.00	0.00	0.07	8.61
	53.48	42.13	3.54	0.03	0.01	0.00	0.81	
	17.90	5.37	5.24	0.81	2.97	0.00	11.77	
1	5,716	17,378	1,260	76	9	1	148	24,588
	17.81	51.16	3.20	0.16	0.01	0.00	0.38	72.72
	24.49	70.35	4.40	0.22	0.02	0.00	0.52	
	69.22	75.79	54.97	47.79	51.19	24.14	63.90	
2	914	3,834	684	50	2	1	38	5,523
	2.50	10.09	1.59	0.10	0.00	0.00	0.10	14.39
	17.39	70.13	11.03	0.71	0.02	0.00	0.72	
	9.73	14.95	27.26	30.16	9.85	24.14	17.52	

3	266 0.61 19.29 2.37	920 2.00 63.08 2.96	251 0.48 15.02 8.17	27 0.06 1.74 16.33	4 0.01 0.23 25.61	1 0.00 0.03 51.72	11 0.02 0.61 3.28	1,480 3.17
4	52 0.13 14.75 0.49	240 0.45 56.54 0.72	92 0.23 26.23 3.89	4 0.00 0.51 1.30	2 0.00 0.23 6.93	0 0.00 0.00 0.00	4 0.02 1.75 2.55	394 0.86
5	18 0.05 30.19 0.19	56 0.09 52.07 0.13	13 0.02 12.00 0.34	2 0.00 1.60 0.78	1 0.00 0.59 3.44	0 0.00 0.00 0.00	2 0.01 3.56 0.99	92 0.16
6	10 0.02 22.88 0.07	21 0.05 56.77 0.07	7 0.01 8.36 0.11	2 0.01 11.99 2.82	0 0.00 0.00 0.00	0 0.00 0.00 0.00	0 0.00 0.00 0.00	40 0.08
7	2 0.00 24.23 0.01	6 0.01 67.65 0.01	1 0.00 8.11 0.01	0 0.00 0.00 0.00	0 0.00 0.00 0.00	0 0.00 0.00 0.00	0 0.00 0.00 0.00	9 0.01
Total	8,262 25.72	23,816 67.50	2,430 5.82	164 0.34	19 0.03	3 0.00	220 0.59	34,914

Table 6-6. Concurrence Between Household and Provider Reports for the General Condition Categories

Category Number	Description	Number of Visits With the Household Reporting the Category
0	No Condition	8,262
1	Infective and Parasitic Diseases	1,166
2	Neoplasms	1,468
3	Endocrine, Nutritional & Metabolic Disorders	1,787
4	Diseases of the Blood & Blood Forming Organs	259
5	Mental Disorders	1,541
6	Nervous System & Sense Organs	2,368
7	Circulatory System	3,982
8	Respiratory System	4,168
9	Digestive System	1,366
10	Genitourinary System	1,729
11	Pregnancy	791
12	Skin & Subcutaneous Tissue	1,351
13	Musculoskeletal System & Connective Tissue	1,553
14	Accidents, Poisonings, & Violence	3,201
15	Paralysis, Congenital Absences, & Impairments	1,152
16	Other Diseases and Conditions	1,377

Weighted Percent of These Visits With Provider Also Reporting the Category	Number of Visits With the Provider Reporting the Category	Weighted Percent of These Visits With Household Also Reporting the Category
18	2,788	53
35	1,271	36
51	1,582	43
66	3,074	36
30	433	19
50	2,240	33
69	3,543	44
68	5,668	48
67	4,869	60
56	2,271	34
70	2,960	42
86	1,747	38
69	2,150	44
63	2,993	35
63	3,024	64
15	465	37
13	4,123	4

general nature of the collapsed set of recodes. It would appear that many individuals did not have even a basic understanding of the nature of the conditions that lead them to seek medical advice. Part of the problem may arise from the fact that some categories are still related. For instance, viral diseases of the conjunctiva are classified into general recode 1 (Infective and Parasitic Diseases). If the individual reported that a visit was for "trouble with my eyes" and his provider said conjunctivitis, the visit would be classified into general recode 6 based upon the household report and into general recode 1 for the provider report. Finally, some of the disagreement between condition reports may have been introduced by the automated matching of household and provider visits. False positive matches induce bias into the estimation of the relationship between the two reports since they result in condition data for different visits being linked.

Some insight was gained concerning the lack of agreement between the two condition reports by examining the relationship between the two sets of condition recodes to see if there was any clustering of the provider diagnoses when the provider failed to agree with the household. This task was complicated by the fact that the two sets of condition recodes are represented by 62 variables which take on the value of zero or one. Obviously, there are many different combinations of condition recodes that could be reported. To simplify the problem and facilitate statistical analysis of the data, this portion of the analysis was restricted to visits where the household reported zero or one condition and the provider also reported zero or one diagnoses. This restriction allowed the definition of two 63-level condition variables, one for the household report and one for the provider report.* Cross tabulations were then made between the household-reported condition and the provider-reported diagnosis. The visits that were used in this analysis constitute 69 percent of all matched visits.

The results of this tabulation show a great deal of clustering of provider diagnostic reports for almost every level of the household condition report variable. To facilitate visual analysis of the tabulation, Table 6-7 was created which gives the principal provider diagnoses associated with each household condition. In creating Table 6-7, the provider diagnostic clusters that are listed are those that occurred with the greatest frequency. Table 6-7 must be examined in conjunction with Table 6-1 to understand the clustering involved.

Some of the conclusions resulting from a close examination of Table 6-7 are as follows. Almost every household report category shows a great deal of clustering of the associated provider reports. In general, the principal provider category associated with a particular household condition category is that same diagnostic category. Commonly occurring as the provider diagnoses over all household-reported categories are recodes 64 (the miscellaneous condition category) and 0 (no condition reported or condition data not provided). The other provider diagnostic reports associated with a

---

*The variables have 63 levels to allow for the 62 condition recodes and the fact that no condition may have been reported.

Table 6-7. The Five Principal Diagnoses Reported by Providers for Each Household-Reported Condition Category

Household-Reported Condition Category	Number of Visits	Principal Provider Diagnoses Associated with the Visits									
		First Category		Second Category		Third Category		Fourth Category		Fifth Category	
		Code	Percent	Code	Percent	Code	Percent	Code	Percent	Code	Percent
0	6,780	64	25	0	21	46	13	19	6	22	3
1	27	1	57	64	19	35	11	48	7	26	2
2	139	35	22	64	15	4	11	20	8	3	6
3	94	3	37	40	15	64	11	35	9	48	7
4	530	4	38	35	12	64	10	48	8	45	6
5	667	5	59	0	15	64	7	48	5	6	4
6	390	6	24	64	18	45	15	5	9	48	8
7	98	7	46	64	17	0	7	45	6	44	5
8	418	8	64	64	12	0	7	52	2	22	2
9	161	9	69	64	16	48	2	35	2	4	1
10	70	10	28	0	23	64	23	42	7	12	5
11	79	0	43	64	20	10	15	5	11	44	4
12	188	0	19	48	15	12	14	3	10	40	9
13	258	15	36	13	31	12	18	0	4	35	3
14	18	14	28	0	26	64	15	35	10	58	9
15	309	13	24	15	19	12	12	0	11	42	6
16	80	16	53	64	14	19	7	56	4	48	3
17	166	17	38	64	7	0	6	13	5	52	5
18	265	18	72	19	17	0	6	64	3	10	1
19	493	19	73	0	6	64	5	48	3	18	3
20	298	20	53	21	18	64	8	35	6	0	3
21	249	21	42	0	18	20	15	35	6	64	6
22	671	22	64	64	7	23	6	0	6	35	2
23	609	23	53	0	12	64	10	35	6	22	3
24	43	0	23	64	18	23	15	59	8	19	7
25	350	25	46	64	7	40	6	0	6	45	4
26	387	35	47	29	8	26	6	64	6	0	6

27	375	35	34	27	9	64	8	3	7	29	5
28	121	28	48	35	17	64	9	29	8	0	4
29	160	29	55	35	44	64	10	0	3	4	3
30	36	30	54	0	32	64	9	35	3	29	2
31	224	48	27	31	26	32	19	0	8	29	8
32	200	32	30	64	22	0	13	35	12	48	10
33	104	33	32	64	18	35	14	26	5	42	4
34	144	34	48	35	31	64	3	4	3	0	3
35	660	35	47	64	9	29	6	34	5	0	4
36	94	36	41	0	15	64	10	40	9	48	6
37	53	64	18	40	15	3	12	35	8	45	8
38	146	38	67	0	7	64	7	40	5	35	5
39	137	39	55	64	13	40	8	0	5	22	4
40	311	40	42	64	13	3	11	0	8	25	4
42	588	42	67	64	7	45	6	0	3	48	3
43	98	43	47	45	12	0	11	64	7	9	3
44	31	43	21	64	18	15	17	45	12	42	8
45	426	45	49	42	11	64	10	43	9	0	6
46	730	46	87	0	7	64	4	45	0+	40	0+
48	983	48	72	64	5	0	5	6	4	4	2
49	337	49	56	52	11	64	9	0	5	54	2
50	71	50	45	51	18	54	9	0	9	49	5
51	63	54	31	51	16	40	7	35	7	22	7
52	382	52	49	64	7	49	7	0	6	62	5
53	650	53	59	0	9	64	8	54	5	56	3
54	279	54	49	64	8	52	6	56	6	53	6
55	458	55	65	0	11	64	4	48	4	56	4
56	209	56	32	53	9	48	8	0	7	58	7
57	82	64	14	0	13	46	11	25	10	57	9
58	470	58	37	48	11	56	9	0	9	64	7
59	43	59	73	0	10	55	9	17	4	6	1
60	132	54	29	51	14	60	14	49	9	64	5
61	85	0	14	60	13	49	13	64	11	55	9
62	224	52	29	62	23	54	11	0	8	64	7
63	210	64	21	9	11	49	9	40	8	0	8
64	826	48	17	64	14	45	10	0	9	22	3

particular household condition report tend to be related to the household report. For instance, when the household reported cataract as the condition (recode 18), the provider agreed 72 percent of the time, reported other conditions of the eye (recode 19) 17 percent of the time, no diagnosis (recode 0) 6 percent of the time, and the miscellaneous category (recode 64) 3 percent of the time. Certain household-reported categories exhibited notoriously poor provider agreement. Examples of this are when the household reported viral infections (recode 2) or diseases of blood and blood forming organs (recode 11). In these cases, the provider principally reported everything but the category that the household reported.

The visual examination of these tables and figures leads one to infer that the household reported condition provides insight into the diagnosis the provider would report but does not always allow accurate prediction of the exact provider diagnosis. Since the main objective of this investigation was to determine the relationship between the household condition report and the provider report, an analysis of the correlation between the two reports appeared to be in order.

By restricting attention to the subgroup of visits with none or one condition, the visits could be cross-classified in terms of the condition the household reported versus the diagnosis reported by the provider with no condition or diagnosis being included as one of the classifications. The resultant table can be analyzed in many different ways to obtain measures of the association between household reported conditions and provider diagnoses. For this study, the most important measures were those that assess the extent of agreement between the two reports.

Because of the provider's access to medical records and his medical training, the provider classification of diagnosis may be viewed as more reliable evaluation of the patient's medical condition than the household classification. Hence, the extent in which the provider classification agrees with the household classification is of interest. In this situation, the probability of agreement is the best measure of association between the two sets of coded data; this probability may be expressed as

$$\sum_{i=1}^{D} p_{ii}$$

where p_{ii} is the weighted percentage of visits for which both the household and the provider reported condition recode i and D is the number of diagnostic categories (D = 63 for original recodes and D = 17 for general recodes). To determine if variations differ across reporting domains, Table 6-8 presents this overall probability of agreement for the total domain and for domains defined by age, race, sex, number of 1977 ambulatory visits, and number of 1977 hospital visits. The overall probability of agreement in this instance is 39 percent when the original condition recodes are used and 48 percent when the generalized recodes are used. That is, 39 percent of the visits have the household and the provider agreeing on the condition category when the original recodes are used. When

Table 6-8. The Overall Probability of Agreement Between Household-Reported Conditions and Provider Diagnoses When the Original Recodes Are Used Versus the Generalized Recodes

Domain		Number of Sample Visits	Percent of Visits in Agreement When Original Recodes Are Used	Percent of Visits in Agreement When General Recodes Are Used
Total		25,739	39	48
Age:	0-4	2,101	38	49
	5-14	2,528	37	52
	15-24	3,918	43	51
	25-34	3,791	39	47
	35-44	2,494	37	47
	45-54	2,854	36	45
	55-64	3,133	38	49
	65+	4,920	38	49
Sex	Male	10,138	40	51
	Female	15,601	38	46
Race	White	22,702	39	49
	Nonwhite	3,037	33	43
Hospital Stays	None	15,397	39	49
	Some	10,342	39	48
Ambulatory Visits	0	65	62	67
	1-5	16,471	39	48
	6-10	4,683	38	48
	11-15	2,067	37	48
	16+	2,453	39	51

the collapsed recodes are used, 48 percent of the visits have household and provider agreeing on the condition.

The most noteworthy result illustrated by Table 6-8 is that there does not appear to be any important variation in the probability of agreement for the various domains. The probability of agreement is marginally higher for whites versus nonwhites and for young adults versus the remainder, but the differences are not large enough to be of practical importance. Individuals with no ambulatory visits do appear to report more accurately; this may be due to the fact that these individuals are reporting the conditions for hospital stays, which may be better known.

Although the overall level of agreement is low, it could be that for some condition categories a higher level of agreement exists. To determine if this is true for the i-th condition, the 63 by 63 table can be collapsed into an i-reported/i-not-reported table and the probability of agreement calculated. Using this method, the probability of agreement for condition i can be expressed as

$$2p_{ii} + 1 - p_{i.} - p_{.i}$$

where the "." notation implies summation over the missing subscript. Table 6-9 presents the probability of agreement for each condition category reported by the household. In general, the level of agreement for each condition is outstandingly high. This should not be particulary surprising, however, since the vast majority of visits are not for a particular condition and are correctly identified as <u>not</u> being for that condition. The health care analyst is much more

Table 6-9. Probability of Agreement Between Household and Provider Report For Each Condition Recode Category

Condition Recode	Percent Agreement	Condition Recode	Percent Agreement
0	71.2	32	98.9
1	99.9	33	99.4
2	99.1	34	99.1
3	99.1	35	94.1
4	97.3	36	99.7
5	98.9	37	99.5
6	98.5	38	99.7
7	99.5	39	99.6
8	99.4	40	98.2
9	98.9	42	98.3
10	99.6	43	99.2
11	99.7	44	99.6
12	98.8	45	97.1
13	98.9	46	95.2
14	99.9	48	95.6
15	98.5	49	98.5
16	99.6	50	99.7
17	99.0	51	99.2
18	99.6	52	97.7
19	97.1	53	98.3
20	98.6	54	98.2
21	98.6	55	98.7
22	97.3	56	98.5
23	97.8	57	99.7
24	99.8	58	97.9
25	98.6	59	99.9
26	98.0	60	99.3
27	98.1	61	99.6
28	99.5	62	99.0
29	98.7	63	99.1
30	99.9	64	84.4
31	99.1		

interested in the conditional probability that the condition is reported correctly for visits where the condition exists. Although these conditional probabilities will never be known, the probability of agreement of the provider with conditions reported by the household can be determined (and vice versa). These probabilities have already been calculated for the full set of visits (see Table 6-3) and in general are rather low considering that the data analyst frequently wants to use condition data to define analysis domains.

6.3. Suggestions for Further Research

The basic conclusion which can be drawn from this study is that there does not appear to be as strong a level of agreement between household and provider reports as the researcher would desire. Part of this lack of agreement might be removed if the definitions for the recode categories were revised. It is obvious that households have difficulty reporting conditions. This implies that the recoding system should be redefined in such a manner that the task of reporting condition accurately is made easier for households or else only provider data should be used for analysis. It would be well worth the effort to investigate modifications of the recoding system that could produce a higher level of agreement between provider and household reports. The level of disagreement between household-reported conditions and provider-reported diagnosis after recoding not only suggests that the recoded household condition data are of questionable value in predicting the exact provider diagnosis but also that the original data (expressed in NHIS Codes) may be even more misleading. This study investigated the issue to a certain extent but there is much more to be done. The data from this investigation would prove valuable in developing a revised coding system to obtain greater agreement between household and provider reports. If the provider diagnosis is desired, then one must conclude that the household data as they stand cannot be used in lieu of obtaining diagnosis data directly from the provider. This leaves the researcher in a quandary since most surveys do not include a provider check. Thus, definitions of condition categories that make it easier for households to provide data consistent with provider diagnosis are of importance to the entire research community.

The results of this investigation of the relationship between household and provider condition reports are both enlightening and disturbing. Considerable insight has been gained into the relationship between the two reports. Unfortunately, this relationship appears to be weak. This investigation suggested that more research is needed to describe the relationship between the two reports and to develop a method for predicting provider diagnosis based upon household reports. Until such research is conducted, the health care researcher would be advised to treat household-reported condition data with caution. The discrepancies between household- and provider-reported data that are described in this chapter should be carefully considered in using household-reported condition data to make inferences about the associated provider diagnosis.

References

Cooley, P. C. and B. G. Cox (1981). An Automated Procedure for Matching Record Check and Household-Reported Health Care Data. Proceedings of the American Statistical Association, Survey Research Methods Section, 418-423.

Cox, B. G. and G. Gridley (1981). The Relationship Between Household and Provider Reported Diagnoses for Medical Care Visits, Paper presented in the American Public Health Association Annual Meetings and available from the National Center for Health Services Research, Rockville, Maryland.

Cox, B. G. and D. S. McGrath (1981). The Relationship Between Household and Provider Reported Diagnoses for Visits Reported in the National Medical Care Expenditure Survey, RTI Report No. RTI/1320-20F, Contract No. HRA 230-76-0268, National Center for Health Services Research, Rockville, Maryland.

National Center for Health Statistics, M. Dicker (1983). Health Care Coverage and Insurance Premiums of Families, United States, 1980, National Medical Care Utilization and Expenditure Survey, Preliminary Data Report No. 3, DHHS Publication No. 83-20000, Washington, DC: U.S. Government Printing Office.

U. S. Department of Health, Education, and Welfare (1968). Eighth Revision International Classification of Diseases, Adapted for Use in the United States, PHS Publication No. 1693, Washington, DC: U. S. Government Printing Office.

U. S. Department of Health, Education, and Welfare (1969). Health Interview Survey: Medical Coding Manual and Short Index, National Center for Health Statistics, Hyattsville, Maryland.

7
Weight Development for Survey Data

Probability sampling is used in surveys so that the sample data can be analyzed to make inferences about the target population of interest. To derive unbiased national estimates of population parameters, the selection probability for each sampling unit must be incorporated into the estimation strategy. This is achieved through the introduction of sampling weights, which serve to differentially weight the sample data to reflect the level of disproportionality in the sample relative to the population of interest. The weight of a sample individual can be viewed as the number of target population members that the sample individual represents. In constructing survey estimates, these sampling weights are applied directly to the data from each member of the sample.

Even when the sample has been designed to allow exact proportional representation of the population of interest, the differential impact of nonresponse and undercoverages leads to a distortion in the sample that requires the construction of differentially adjusted weights for use in analysis. For instance, surveys are burdened to differing degrees by nonresponse of eligible sample members. Nonresponse adjustments to the sampling weights are often used to reduce the bias induced in survey estimates through the loss of data for nonrespondents. Post-stratification adjustments are often made to improve the precision of sample estimates by using more accurate ancillary sources of information for population totals. Upper bounds are occasionally placed on the magnitude of resultant weights to reduce the variance inflating effect that excessively large weights has for survey estimates.

In this chapter, the methodology underlying the construction of sampling weights is presented, together with a description of adjustments that may be made to account for sample nonresponse and undercoverage. The statistical theory motivating the use of weighting strategies to adjust for sample nonresponse is presented and the effect of excessively large weights is then described. The National Medical Care Expenditure Survey (NMCES) is similar in design to many national surveys, for example, the National Medical Care Utilization and Expenditure Survey (NMCUES), the National Health Interview Survey (NHIS), and the National Health and Nutrition Examination Survey (NHANES). As an illustration of the weight development process for survey data, NMCES weighting is described in the conclusion of this chapter (Cohen and Kalsbeek, 1981).

7.1 Initial Sampling Weights

Stratified, multi-stage, area probability samples allow approximately unbiased estimation of health parameters at the national level. This is conditioned upon the application of weights to the sample data that properly reflect the sample selection scheme. The <u>sampling weight</u> for a sample member is defined as the inverse of the probability of selection for without replacement sampling and as the inverse of the expected frequency of selection for with replacement sampling. For nested sample designs, the weights will be the inverse of the product of the sampling weight components associated with selection at each stage. In a five-stage sample design similar to the NMCES, the initial sampling weight for the e-th member of the d-th dwelling unit in the c-th area segment from the b-th second-stage unit in the a-th primary sampling unit is defined as

$$\begin{aligned} w(abcde) &= [\Pi(abcde)]^{-1} \\ &= [\Pi(a)\cdot\Pi(b|a)\cdot\Pi(c|ab)\cdot\Pi(d|abc)\cdot\pi(e|abcd)]^{-1} \qquad (7\text{-}1) \end{aligned}$$

where

- $w(abcde)$ is the sampling weight for the e-th member of the d-th dwelling unit in the c-th segment from the b-th second-stage unit in the a-th primary sampling unit,
- $\Pi(abcde)$ is the selection probability for the e-th member of the d-th dwelling unit in the c-th area segment of the b-th secondary sampling unit from the a-th primary sampling unit,
- $\Pi(a)$ is the first-stage probability of selecting the a-th primary sampling unit,
- $\Pi(b|a)$ is the second-stage conditional probability of selecting the b-th secondary sampling unit given the a-th primary sampling unit is selected,
- $\Pi(c|ab)$ is the third-stage conditional probability of selecting the c-th area segment given the b-th secondary sampling unit of the a-th primary sampling unit is selected,

$\Pi(d|abc)$ is the fourth-stage conditional probability of selecting the d-th dwelling unit given the c-th segment of the b-th secondary sampling unit of the a-th primary sampling unit is selected, and

$\pi(e|abcd)$ is the fifth-stage conditional probability of selecting the e-th individual given that the d-th dwelling unit of the c-th segment of the b-th secondary sampling unit of the a-th primary sampling is selected.

These sampling weights act as inflation factors to represent the number of units in the survey population that are accounted for by the sample unit to which the weight is assigned. Their sum provides an unbiased estimate of the total number of individuals in the target population.

To compensate for the bias induced by survey nonresponse, the sampling weights of the respondents are adjusted so that they sum to the population total. Common practice is to first partition sample members into groups related to response to study variables and then to adjust the weights of the respondents within each group so that they sum to the population total for the group. These adjusted weights are referred to as analysis weights. When the population totals are known (e.g., frame counts or census information), the adjustment process is referred to as post-stratification and the groups as post-strata. When population totals are unknown, the totals are estimated from the sample data itself. This process is referred to as weighting class adjustment and the groups as weighting classes. Occasionally this process produces large variations in the adjusted weights that may have deleterious effects on the variances of sample estimates. In this situation, the nonresponse adjustment factors or the weights may be truncated to reduce the range of values for the final analysis weights.

7.2 Impact of Survey Nonresponse

Survey nonresponse is characterized by an inability to obtain information for all sample members. For most surveys, nonresponse involving the entire interview or questionnaire can be classified into four distinct categories: noncoverage, not at home, unable to answer and hard-core nonrespondents (Cochran, 1977, p. 364). Noncoverage is defined as a failure to find or make contact with units selected for the sample. Persons who reside at home but are temporarily away from the house during the times interviewers call constitute the not at homes. The unable to answer group are characterized by respondents not in possession of the information requested, or unwilling to obtain it. Hard core nonrespondents include individuals who consistently refuse to be interviewed, who are incapacitated, or who are away from home during the entire survey period.

Reporting of the levels of nonresponse experienced in a sample survey is essential to ascertain the representativeness of the survey data. Accurate counts of the response status of all eligible sample members are required for correct measurement of the overall nonresponse rate. Nonresponse is widely viewed as a potentially important source of error or bias in survey estimates. These effects become more pronounced with increasing levels of nonresponse and with substantial differentials in data profiles between respondents and nonrespondents.

The impact of complete nonresponse on estimation is illustrated in the following example. Suppose that an estimate is desired of the mean response to a criterion variable of interest for the target population of a study. The parameter for which an estimate is desired is

$$\bar{Y} = \sum_{i=1}^{N} Y(i)/N \qquad (7\text{-}2)$$

where Y(i) is the criterion variable response of the i-th population member and N is the population size. If complete responses were obtained for all sample selections, then an unbiased estimate of the population mean $\bar{Y}$ would be provided by the sample mean $\bar{y}$ or

$$\bar{y} = \sum_{i \varepsilon S} w(i)\, Y(i) \Big/ \sum_{i \varepsilon S} w(i) \qquad (7\text{-}3)$$

where w(i) is the sampling weight and $i \varepsilon S$ denotes that the summation is over all members of the sample. However, 100 percent response is almost never obtained for sample selections. Consider instead the use of questionnaire responses from survey respondents to estimate the mean for the total population. That is, $\bar{Y}$ would be estimated by $\bar{y}_r$ where

$$\bar{y}_r = \sum_{i \varepsilon SR} w(i)\, Y(i) \Big/ \sum_{i \varepsilon SR} w(i) \qquad (7\text{-}4)$$

and $i \varepsilon SR$ indicates that the summation is over all sample respondents. The bias of $\bar{y}_r$ as an estimate of the population mean $\bar{Y}$ is defined as the expected value over repeated sampling of the difference between $\bar{y}_r$ and $\bar{Y}$ or

$$\text{Bias}(\bar{y}_r) = E(\bar{y}_r - \bar{Y}) = \bar{Y}_r - \bar{Y} \qquad (7\text{-}5)$$

where $\bar{Y}_r$ is the mean response of those population members that would respond if selected for the survey. Note that the population mean $\bar{Y}$ can be expressed as:

$$\bar{Y} = P_r \bar{Y}_r + P_{nr} \bar{Y}_{nr} \tag{7-6}$$

where

P_r is the proportion of the N population units that would respond if selected for the survey,

P_{nr} is the proportion of the N population units that would not respond if selected for the survey, and

$\bar{Y}_{nr}$ is the mean of the criterion variable for all those who would not respond if selected.

Noting that $P_{nr} = 1 - P_r$, the bias expression in equation (7-5) can be re-expressed as:

$$\text{Bias}(\bar{y}_r) = P_{nr} (\bar{Y}_r - \bar{Y}_{nr}) \tag{7-7}$$

Thus, bias of the respondent mean as an estimate of the population mean will vary depending upon the rate of nonresponse and the difference between the respondent population mean and the nonrespondent population mean (Chapman, 1976; Kish, 1965, pp. 535-536).

A similar situation will occur when the respondent total is used to estimate the population total. The population parameter of interest here is Y where

$$Y = \sum_{i=1}^{N} Y(i) \tag{7-8}$$

The estimated total from the respondent data takes the form

$$\hat{Y}_r = \sum_{i \varepsilon SR} w(i)\, Y(i) \tag{7-9}$$

The bias of the respondent total as an estimate of the population total is

$$\text{Bias}\ (\hat{Y}_r) = E(\hat{Y}_r - Y) = Y_r - Y \tag{7-10}$$

where Y_r is the total across all members of the population who would respond if included in the survey. Noting that

$$Y = Y_r + Y_{nr} \tag{7-11}$$

where Y_{nr} is the total for the nonresponding population, the following alternative expression for the bias is obtained:

$$\text{Bias}\ (\hat{Y}_r) = -\ Y_{nr} \tag{7-12}$$

That is, the respondent total will be negatively biased by an amount equal to the nonrespondent population total. Thus, when the respondent population mean is exactly equal to the nonrespondent population mean, the respondent sample mean will be an unbiased estimate of the population mean but the respondent total will not be unbiased for the population total.

To compensate for the bias induced by survey nonresponse, nonresponse adjustments may be made to the sampling weights to inflate the responding sample to represent the total sample, under the assumption that the responding subset is representative of the total.

Consider for the moment, an overall nonresponse adjustment applied to the sampling weights to force the respondents' weights up to the total population estimate obtained from the sample. Define the adjustment factor:

$$A_0 = \sum_{i \varepsilon S} w(i) \Big/ \sum_{i \varepsilon SR} w(i) \tag{7-13}$$

where w(i) is again the sampling weight for the i-th sample member, $i\varepsilon S$ indicates summation over the full sample, and $i\varepsilon SR$ indicates summation over the sample respondents. The overall nonresponse adjusted weight $[W_0(i)]$ would then be computed as the product of the adjustment factor and the sampling weight or:

$$W_0(i) = A_0\, w(i) \tag{7-14}$$

This overall weighted adjustment for nonresponse is incorporated in the estimation of a population mean for a criterion variable of interest in the following manner:

$$\bar{y}_o = \sum_{i \varepsilon SR} W_0(i)\, Y(i) \Big/ \sum_{i \varepsilon SR} W_0(i) \tag{7-15}$$

As in the previous section, the bias of $\bar{y}_o$ reduces to

$$\text{Bias}\ (\bar{y}_o) = (1-P_r)\ (\bar{Y}_r - \bar{Y}_{nr}) \tag{7-16}$$

which is a function of the population nonresponse rate, and the difference between the population means for respondents and nonrespondents. Hence, an overall weight adjustment does not reduce the bias of estimated population means.

A population total for a criterion variable of interest would be computed as:

$$\hat{Y}_o = \sum_{i \varepsilon SR} W_0(i)\, Y(i) \tag{7-17}$$

Note that the expected value of $\hat{Y}_o$ over repeated samplings will be $N\,\bar{Y}_r$. The population total that is being estimated can be expressed as

$$Y = N_r\,\bar{Y}_r + N_{nr}\,\bar{Y}_{nr} \tag{7-18}$$

where N_r and N_{nr} are the population counts of respondents and nonrespondents, respectively. The bias of the adjusted population total $\hat{Y}_o$ will be

$$\text{Bias}\,(\hat{Y}_o) = N\,\bar{Y}_r - (N_r\,\bar{Y}_r + N_{nr}\,\bar{Y}_{nr}) \tag{7-19}$$

or

$$\text{Bias}\,(\hat{Y}_o) = N_{nr}\,(\bar{Y}_r - \bar{Y}_{nr}) \tag{7-20}$$

When the respondent mean is equal to the nonrespondent mean, $\hat{Y}_o$ will be an unbiased estimate of Y. In this situation, an overall weight adjustment will reduce the bias of estimated population totals.

7.3 Weighting Class Adjustments for Total Nonresponse

Properly designed, a weighting class nonresponse adjustment strategy will result in a reduction in nonresponse bias. The technique requires a partitioning of the population into mutually exclusive classes, with membership defined by classification information available for both responding and nonresponding units. This requirement restricts the set of potential classification variables. Consequently, the classes are often defined by geographic or other sample design parameters. The specification of the weighting classes should be governed by an attempt to combine sampling units into internally homogeneous groups with respect to criterion measures of greatest interest to the survey.

To illustrate the procedure, assume the sample has been divided into C weighting classes. Let w(ci) represent the sampling weight for the i-th sampling unit in the c-th class. Similar to the overall weighted nonresponse adjustment, the sum of the respondents' weights are weighted up to the total sample estimate within each of the C specified classes. The adjustment for the c-th class (c=1,2,...,C) is computed as:

$$A(c) = \sum_{i \varepsilon c} w(ci) \Big/ \sum_{i \varepsilon cr} w(ci) \tag{7-21}$$

where $i\varepsilon c$ indicates summation over all sample members in the c-th weighting class and $i\varepsilon cr$ indicates summation over all respondents in the c-th weighting class. The nonresponse adjusted weight of survey respondents is then obtained as the product of the adjustment factor and the sampling weight or

$$W_a(ci) = A(c)\ w(ci) \tag{7-22}$$

An estimate of the population mean, $\bar{Y}$, for a criterion variable of interest would be expressed as

$$\bar{y}_a = \sum_{c=1}^{C} \sum_{i \varepsilon cr} W_a(ci)\ Y(ci) \Big/ \sum_{c=1}^{C} \sum_{i \varepsilon cr} W_a(ci) \tag{7-23}$$

The bias of $\bar{y}_a$, the sample estimate of the population mean that incorporates the weighting class nonresponse adjustment, is approximately (Chapman, 1976):

$$\text{Bias}(\bar{y}_a) \doteq \sum_{c=1}^{C} P(c)\,[P_{nr}(c)]\,[\bar{Y}_r(c) - \bar{Y}_{nr}(c)] \qquad (7\text{-}24)$$

where

$P(c)$ is the proportion of the total population in the c-th weighting class,

$P_{nr}(c)$ is the proportion of population units in the c-th weighting class that would not respond if selected for the survey,

$\bar{Y}_r(c)$ is the criterion variable mean for the population members of the c-th weighting class who would respond if selected, and

$\bar{Y}_{nr}(c)$ is the criterion variable mean for the population members of the c-th weighting class who would not respond if selected.

When the nonresponse rates, $P_{nr}(c)$, vary across the C classes and the absolute differences in criterion variable means are generally smaller within weighting classes, $[\bar{Y}_r(c) - \bar{Y}_{nr}(c)]$, than the absolute difference in the overall criterion variable means, $|\bar{Y}_r - \bar{Y}_{nr}|$, the use of a weighting class nonresponse adjustment will result in a reduction of nonresponse bias. A similar situation will occur for estimated population totals.

The precision of survey estimates derived with weighting class nonresponse adjustments is affected by the number of sample respondents in each class. The variance of an estimated mean will also increase with the application of large adjustment factors, which are necessary to accomodate those classes with a small number of sample respondents relative to the class total. To reduce the adverse effects of relatively large nonresponse adjustment factors on the precision of survey estimates, an upper bound is often placed on their order of magnitude. In many of the national surveys, (which include the NMCES and NMCUES), the upper bound is specified as two. Weighting classes are often collapsed to satisfy this constraint. An additional requirement of a minimum of 20 sample respondents per weighting class is often invoked to insure an adequate level of sample representation.

7.4 Post-Stratification Adjustment for Total Nonresponse and Undercoverage

Use of a stratified sampling design is characterized by gains in precision for survey estimates, particularly when the strata are homogeneous with respect to criterion variables of interest. Prior to selection, however, it may be inconvenient or impossible to classify all eligible sampling units into analytically desirable strata. To improve upon the precision of survey estimates, post-stratification or stratification after selection is often employed to complement the original stratification scheme. Another reason for post-stratification occurs when information on the stratification variable(s) is not available prior to sample selection. Occasionally, this information is available, but the costs in dollars and time of stratifying every population member into strata are prohibitive. In addition, relevant stratification variables are sometimes overlooked at the time of selection.

Post-stratification adjustment requires knowledge of the proportion of the population belonging to the specified strata, and data to classify respondents into the respective strata (Kish, 1965 pp. 90-92). The <u>post-stratification adjustment</u> forces the sampling weights within each post-stratum to the known population total for the post-stratum. The gains in precision from post-stratification can approach those obtained from a stratification scheme with proportionate allocation (Cochran, 1977, pp. 134-135). Post-stratification also serves as a correction for nonresponse and undercoverage error. It is the preferred alternative to a weighting class adjustment which only focuses on nonresponse bias. In national surveys, undercoverage is primarily due to sampling frame deficiencies, where the available list of sampling units does not identify all the eligible target population members. The relative reduction in bias that is achieved through post-stratification is often of the same magnitude as the corresponding decrease in variance.

The post-stratification adjustment of sampling weights and their use in survey estimation is as follows. Let N(h) represent the number of population units in the h-th post-stratum, where h=1,2,... H, and H is the total number of post-strata. The post-stratification adjustment to the sampling weight, w(hi), of the i-th sampling unit classified in the h-th post-stratum is computed as:

$$A(h) = N(h) \; / \sum_{i \varepsilon hr} w(hi) \qquad (7\text{-}25)$$

where iεhr indicates summation over all respondents in the h-th stratum and w(hi) is the sampling weight. The post-stratified adjusted weight is expressed as:

$$W_s(hi) = A(h)\ w(hi) \tag{7-26}$$

The post-stratified survey estimator for the criterion variable of interest takes the form:

$$\bar{y}_s = \sum_{h=1}^{H} \sum_{i \varepsilon hr} W_s(hi)\ Y(hi) \ / \ \sum_{h=1}^{H} \sum_{i \varepsilon hr} W_s(hi) \tag{7-27}$$

where Y(hi) represents the criterion variable measurement for the i-th sample unit in the h-th post-stratum.

7.5 Weight Truncation

Once the sampling weights have been adjusted for nonresponse, an examination of the distribution of the adjusted weights is recommended. Excessively large variation in the analysis weights has a negative effect on the precision of survey estimates. Often, very large analysis weights are truncated by placing an upper bound on their value, and the entire set of analysis weights further adjusted to sum to the weighted total of the original untruncated weights. The truncated and adjusted analysis weights are derived as

$$W_t(i) = T\ \mathrm{Min}[W(i), W_u] \tag{7-28}$$

where T is the sum of the untruncated weights divided by the sum of the truncated weights or

$$T = \sum_{i \varepsilon SR} W(i) \ / \ \sum_{i \varepsilon SR} \mathrm{Min}[W(i), W_u] \tag{7-29}$$

W_u is the maximum allowable value for the analysis weights and W(i) is the analysis weight for the i-th sample unit. $Min[W(i),W_u]$ takes the minimum value of the two weights specified. This process is referred to as weight truncation or smoothing. When this technique is used, the final survey estimates may be biased. However, the associated variance may also be reduced. When excessively large analysis weights are truncated, the achieved decrease in variance may more than offset the increase of bias incurred, yielding a smaller mean square error for survey estimates.

Weight construction is often perceived as a complex operation in part due to the many steps needed to produce analysis weights that reflect the differential probabilities of selection while properly accounting for the effects of nonresponse and under-coverage. To facilitate understanding of the weight construction process, the steps involved in creating analysis weights for the NMCES will be described in the remainder of this chapter. A general discussion of the weighting strategy and its rationale will be presented first and then the actual formulas used in constructing the weight components at each step.

7.6 Estimation in the NMCES

In the NMCES, the dwelling unit constituted the ultimate sampling unit. A dwelling unit was defined as a house, apartment, group of rooms, or a single room that was occupied (or vacant but intended for occupancy) as separate living quarters. Although dwelling units were originally established as the sampling units in the NMCES, the analytical goals of the study necessitated following both families and individuals over the course of 1977 via household reporting units. A reporting unit was established to take account of the fact that dwelling units may contain unrelated persons, while for purposes of NMCES data analysis families had to be identified. In general, sample dwelling units consisted of one or more reporting units with each reporting unit composed of individuals related to one another by blood, marriage, or adoption. Any person 17 years or older who was unrelated to other persons in the dwelling unit of the individual's usual residence was considered as a separate reporting unit.

Since the reporting units were subunits of the dwelling units and all those identified were selected in the sample, the reporting units were assigned the sampling weight of the dwelling unit from which they originated. These reporting unit weights were further modified to produce an individual's analysis weight adjusted for nonresponse, under-coverage, and other factors. The adjustments were computed separately for each quarter of 1977 and for the entire year. This allowed for the derivation of national estimates of health services, utilization, expenditures, and insurance for the U.S. population in each quarter of 1977 and the entire year.

7.6.1 Nonresponse Adjustment in the NMCES

The response rate for the first round of data collection was 91 percent. Nonresponse adjustments at the reporting unit level were computed separately for the RTI and NORC half-samples and incorporated into the analysis weight. The adjustment was applied to account for total nonresponse of entire reporting units. This nonresponse adjustment was made within the strata used in selecting the secondary sampling units. Specifications restricted this adjustment to be no greater than two to limit its effect on the variability of the final analysis weights.

Computation of this adjustment required the determination of eligibility of each unit for the NMCES. Reporting units were ineligible if the listed dwelling unit associated with that reporting unit was vacant, had been demolished, had moved, did not satisfy the definition of a dwelling unit, was under construction, or had merged with another listed dwelling unit. Also ineligible were reporting units found in dwelling units used solely as a vacation home, those where all members were serving in the armed forces or were institutionalized, and those whose members were college students considered to be a part of their parents' reporting unit.

For all eligible reporting units, a second determination as to response status was required. Nonresponding reporting units were defined as:

- A dwelling unit determined to be occupied but where no contact was made with any of the household members.
- A dwelling unit where no contact was ever made with a member at least 14 years of age.
- A reporting unit where contact was established with an eligible respondent who refused to be interviewed or where an insurmountable language barrier existed.
- A reporting unit where the screening information had been obtained but the members subsequently refused to respond to the questionnaire.

An additional nonresponse adjustment was implemented to account for the fact that some initially responding persons did not provide data for the entire year or for the entire time period during which they were eligible to respond. Reasons for this type of nonresponse were:

- Persons who cooperated in the first interview would occasionally refuse to participate in the remainder of the interviews.

- In some cases, contact could not be made with persons who changed residence.

The adopted nonresponse adjustment strategy for partial nonresponse is presented in Chapter 10.

7.6.2 Smoothing Adjustment in the NMCES

To reduce the possibility of a few excessively large weights dominating any particular estimate or inflating the variance of the estimate, a smoothing adjustment was applied. This procedure placed an upper bound on the magnitude of the analysis weights, defined for NMCES as the 98th percentile of the distribution of sampling weights. Again, the adjustment was made separately within survey organizations.

7.6.3 Post-Stratification Adjustment in the NMCES

The weight for each survey respondent was further adjusted to an independent estimate for the national population within each of 32 post-strata defined by the three way cross-classification of age (0-4, 5-14, 15-24, 25-34, 35-44, 45-54, 55-64, 65 and up), race (white, nonwhite), and sex (male, female) characteristics. The purpose of this post-stratification adjustment was to reduce the bias due to nonresponse and to achieve much of the gain in precision that could have been attained had the sample initially been drawn from the population stratified by age, race, and sex. The post-stratification adjustment was made separately for the samples from each survey organization. Independent estimates of the 32 post-strata population totals were taken from Census Bureau projections of the population for the respective quarter of 1977.

7.6.4 Adjustment for Combining Samples

Direct application of the sampling weights so adjusted would yield independent estimates of population parameters for the RTI and NORC half-samples. Consequently, a constant adjustment factor of 0.5 was applied to each respondent's weight to account for the fact that data were obtained from two independent samples of approximately the same size. By computing the arithmetic mean of these two independent estimates, national estimates of population parameters result when both samples are combined.

7.7 Estimation Equations in the NMCES

A general estimation equation and detailed computational formulas for respective adjustments to the sampling weights in the NMCES follow.

7.7.1 General Estimation Equation

The mean estimator of a health-related characteristic of interest takes the form:

$$\bar{y} = \frac{\sum_{g} \sum_{h \varepsilon q} \sum_{a} \sum_{b} \sum_{c} \sum_{d} \sum_{e \varepsilon g} W(habcde)Y(habcde)}{\sum_{g} \sum_{h \varepsilon q} \sum_{a} \sum_{b} \sum_{c} \sum_{d} \sum_{e \varepsilon g} W(habcde)} \tag{7-30}$$

where

W(habcde) is the analysis weight for the e-th individual from the d-th responding reporting unit in the c-th segment, b-th secondary sampling unit, a-th primary sampling unit from the h-th primary stratum. The final analysis weight will reflect nonresponse, smoothing, and post-stratification adjustments.

Y(habcde) is a health characteristic measure for the e-th individual belonging to the g-th weighting class from the d-th responding reporting unit in segment habc.

g is a weighting class indicator defined by a three-way cross-classification of age, race, and sex (g=1,2,...,32).

q is the respective survey organization.

ε indicates membership in q or g.

7.7.2 Nonresponse Adjusted Weight

The reporting unit nonresponse-adjusted sampling weight takes the form:

$$W_1(habcd) = A(hab)\ w(habcd) \tag{7-31}$$

where

A(hab) is the reporting unit nonresponse-adjustment factor at the secondary sampling unit level, and

w(habcd) is the sampling weight for the d-th responding reporting unit in the c-th segment, b-th secondary sampling unit, a-th primary sampling unit from the h-th stratum.

The final reporting unit nonresponse-adjustment factor is determined by

$$A(hab) = B(q)\ \mathrm{Min}[A'(hab),2] \tag{7-32}$$

where

$$A'(hab) = \frac{\sum_{d \varepsilon E(hab)} w(habcd)}{\sum_{d \varepsilon R(hab)} w(habcd)}, \tag{7-33}$$

$$B(q) = \frac{\sum_{h \varepsilon q} \sum_a \sum_b \sum_c \sum_d w(habcd) A'(hab)}{\sum_{h \varepsilon q} \sum_a \sum_b \sum_c \sum_d w(habcd) \text{ Min}[A'(hab), 2]} \tag{7-34}$$

E(hab) is the set of eligible reporting units in secondary sampling unit hab,

R(hab) is the set of eligible reporting units responding in secondary sampling unit hab, and

Min[A'(hab), 2] is a function which takes the minimum value of the two specified quantities.

Note that the estimation procedure provides independent weights for each survey organization q.

7.7.3 Smoothing Adjustment

Since all eligible individuals in a reporting unit were selected for the sample, each individual in a responding reporting unit was given the reporting unit's nonresponse-adjusted analysis weight, or

$$W_1(habcde) = W_1(habcd) \tag{7-35}$$

The smoothed weight adjusted for nonresponse takes the form:

$$W_2(habcde) = T_q(g) \text{ Min } [W_1(habcde), W_1^{\alpha}(habcde)] \tag{7-36}$$

where

$$T_q(g) = \frac{\sum_{h \varepsilon q} \sum_a \sum_b \sum_c \sum_d \sum_{e \varepsilon g} W_1(habcde)}{\sum_{h \varepsilon q} \sum_a \sum_b \sum_c \sum_d \sum_{e \varepsilon g} \text{Min } [W_1(habcde), W_1^{\alpha}(habcde)]} \tag{7-37}$$

$W_1^{\alpha}(habcde)$ is the maximum allowable value for the weights, determined by the 98th percentile of the nonresponse adjusted weights, and

Min $[W_1(habcde), W_1^{\alpha}(habcde)]$ takes the minimum value of the weights specified.

7.7.4 Post-Stratification

The nonresponse-adjusted, smoothed, and post-stratified weight takes the form:

$$W_3(habcde) = W_2(habcde)\ R_q(g) \tag{7-38}$$

for $e \varepsilon g$, where

$$R_q(g) = C_q(g) \Big/ \sum_{h \varepsilon q} \sum_a \sum_b \sum_c \sum_d \sum_{e \varepsilon g} W_2(habcde) \tag{7-39}$$

and $C_q(g)$ is an independent estimate of the noninstitutionalized U.S. population, as provided by the Bureau of the Census for the g-th post-stratum during the respective quarter of 1977 or the entire year.

7.7.5 Adjustment for Combining Samples

Direct application of the weights to data at the individual level will provide independent, approximately unbiased estimates of population parameters using either the RTI or NORC half-sample. Consequently, applying the factor 0.5 to the analysis weights allows for approximately unbiased estimates of the respective population parameters when the two samples are combined. Hence,

$$W_4(habcde) = 0.5\ W_3(habcde) \tag{7-40}$$

constitutes the final NMCES analysis weights.

References

Chapman, D. W. (1976). A Survey of Nonresponse Imputation Procedures. Proceedings of the American Statistical Association, Social Statistics Section, 245-259.

Cochran, W. G. (1977). Sampling Techniques, New York: John Wiley and Sons.

Cohen, S. B. and W. D. Kalsbeek (1981). NMCES Estimation and Sampling Variances in the Household Survey. Instruments and Procedures 2, DHHS Publication No. 81-3281, Washington, DC: U.S. Government Printing Office.

Kish, L. (1965). Survey Sampling, New York: John Wiley and Sons.

8
Imputation Procedures to Compensate for Missing Responses to Data Items

Nonresponse in sample surveys occurs at many levels. The entire set of survey data is lost when an individual refuses to participate in the study or cannot be found at home after repeated callbacks. This type of nonresponse is commonly referred to as total nonresponse. When the survey involves multiple instruments or multiple rounds of data collection, partial nonresponse may occur in which some individuals initially respond but then fail to complete all of the required instruments or rounds of data collection. Within otherwise completed questionnaires, certain question items may have missing responses due to refusal or insufficient knowledge on the respondent's part. This type of missing data, referred to as item nonresponse, will also occur when invalid data or data that violate skip patterns are discarded.

If the answers for most data items were distributed in a similar manner for respondents and nonrespondents, there would be no need to be concerned about the occurrence of missing data. In this situation, the available data could be analyzed directly and reliable estimates obtained for most types of survey statistics. However, the distribution of respondent data is often different from the nonrespondent distribution; hence, survey estimates obtained from respondent data will be biased with respect to describing characteristics of the survey population unless compensations are made for the missing data. For this reason, procedures must be implemented during data processing that reduce the bias caused by missing data.

As discussed in Chapter 7, a sample weighting adjustment is probably the most appropriate solution for total nonresponse. Using this approach, the sampling weights of the respondents are differentially adjusted so that the reweighted respondent distribution for survey questions approximates the response distribution that would have been obtained from the total sample.

When the proportion of respondents providing partial data is small, a cost effective strategy may be to treat partial nonrespondents as total nonrespondents and develop nonresponse adjusted weights only for those individuals with complete data. Chapter 10 and 11 discuss this reweighting approach and alternative data replacement strategies for partial nonresponse in a longitudinal survey context.

With respect to missing item data within an otherwise completed questionnaire, information exists for the nonrespondent that can be used to predict the missing response. Since few persons are able to answer all items on survey questionnaires, completely discarding the data for item nonrespondents is not feasible. Instead, researchers usually replace missing item data by imputed values (Chapman, 1976). This chapter describes alternative item nonresponse imputation procedures and suggests circumstances for which the procedures are appropriate.

8.1 Response Error Bias in Survey Data

Estimates of population values obtained from survey data are subject to two types of error. Variable error is the random component that includes sampling error (due to the random selection of individuals rather than a complete census) and variable measurement errors (due to natural fluctuations in questionnaire responses and data transcriptions). The second error type is bias, systematic errors in survey estimates which result from the estimation procedure, survey nonresponse, or nonsampling errors inherent in the measurement process.

The total error associated with using the sample estimate $\hat{\theta}$ to predict the population parameter θ is the difference between the estimator and the parameter or $\hat{\theta} - \theta$. Let $E(\hat{\theta})$ represent the average or expected value of the statistic $\hat{\theta}$ over repeated samples and repeated transcriptions of each set of sample data. Using this notation, the total error may be modeled as

$$\hat{\theta} - \theta = [\ \hat{\theta} - E(\hat{\theta})\ ] + [\ E(\hat{\theta}) - \theta\ ] \tag{8-1}$$

The first term, $[\ \hat{\theta} - E(\hat{\theta})\ ]$, represents the variable errors associated with the sampling and measurement process. The second term,

$[E(\hat{\theta}) - \theta]$, represents bias, the deviation of the sample statistic over repeated trials from the true population value.

The ultimate goal for the survey planner is to design a sample survey in which the total error of survey results is minimized (not just the sampling error). This implies that attention must be paid to the protocol for collecting the data, the design of the survey instrument, and provisions to minimize nonresponse. Similarly, in using the survey data, the investigator must also be concerned with minimizing the total error. After the data are collected, however, the variable portion of the error is usually fixed. In this case, the investigator should examine the extent of the various kinds of nonresponse and the amount of systematic measurement error in survey responses. Imputation and weighting procedures are available to reduce the bias caused by nonresponse, and logical editing may be used to resolve inconsistencies that are symptomatic of response errors in the data.

When imputation is used to replace missing data, the variance of survey estimates will be greater than if no imputation had occurred. Thus, in choosing an imputation procedure to replace missing data, the researcher must consider the effect on the variability of the survey estimate as well as the bias. The best imputation procedure is one that produces the most accurate survey estimates.

A useful criterion for accuracy is the mean square error. The mean square error of an estimator $\hat{\theta}$ is defined as the expected value of the squared total error or

$$MSE(\hat{\theta}) = E(\theta - \hat{\theta})^2 \tag{8-2}$$

Note that the mean square error of $\hat{\theta}$ may be rewritten as

$$MSE(\hat{\theta}) = E[\hat{\theta} - E(\hat{\theta})]^2 + [E(\hat{\theta}) - \theta]^2 \tag{8-3}$$

or

$$MSE(\hat{\theta}) = Var(\hat{\theta}) + [Bias(\hat{\theta})]^2 \tag{8-4}$$

The first term of the mean square error is the variance of the estimator $\hat{\theta}$; the second term is the squared bias of the estimator. An imputation procedure is effective when it minimizes the mean square error of the survey estimates that will be produced using the imputation-revised data.

8.2 The Role of Imputation in Data Analysis

In performing any data analysis, the researcher makes assumptions about the characteristics of nonrespondents and how their missing data will impact on survey estimates. These assumptions may be explicitly made or they may instead be the implicit basis of the estimation procedure that the researcher selects. Depending upon the extent to which data are missing from the analysis, the method selected by the researcher to deal with missing data can have a significant impact on the accuracy of survey estimates. Some of the methods that have been used in the past include:

- the no imputation approach which ignores the missing data,
- logical imputation which replaces the missing values using the responses to similar data items as a guide,
- mean value imputation which replaces missing data with the average of the respondent data,

- cold deck imputation which uses data from other sources to replace missing values,
- hot deck imputation which uses data from survey respondents to replace missing values, and
- regression imputation which uses respondent data to model question response in terms of other data item responses and then replaces missing data with the model prediction.

In order to select an appropriate procedure and apply it in an effective manner, the researcher must understand the nature of the missing data, that is, the patterns of nonresponse to the survey. For instance, many techniques require that the analyst divide the sample into groups in such a manner that _within each group_ the responses for nonrespondents (if they had been obtained) are similar to those of the respondents. In this situation, the researcher must apply his knowledge of the subject matter and information provided by the respondent data to model the nature of data missingness.

Typically, the researcher cannot validate his model for imputing missing data, since validation requires knowledge of the responses that the nonrespondents would have given had they provided data. However, respondent data can be used to provide insight into variables that are important with respect to modeling nonresponse. For instance, it is well known that blacks tend to have higher unemployment rates than whites and that response rates for blacks tend to be lower than for whites. In modeling the nature of missingness, it is clear that race must be considered for an employment survey. But is this also true for a health care survey? Survey data can be analyzed to answer this question by comparing response rates for whites versus blacks and health care statistics estimated from respondent data. Thus, the survey data can be used to derive a model that describes the missingness in the data.

The following sections of this chapter discuss logical and statistical imputation procedures that can be used to replace missing data. Logical imputation deduces the response that is missing using the responses to other data items as a guide. The statistical imputation procedures presume a model for the missing data which specifies that respondents and nonrespondents have similar responses after variations in the two response groups with respect to related variables have been controlled. Three basic model types are assumed by these procedures: the distance function model, the missing at random model, and the regression model.

Perhaps the most common model for the nature of missingness is the _distance function model_. Within specified groups or imputation classes, the distance function model hypothesizes that missing data can be explained in terms of a distance function associated with a quantitative variable. For instance, if missing hourly wage data are being imputed, the researcher might believe that within occupational groups the missing hourly wage will tend to be similar to that of other individuals with an equal amount of seniority. That is, if the person records within occupation groups were ordered by the number of years of seniority, the researcher would be hypothesizing that records close together in the data file would have similar responses.

Perhaps the second most common model for the nature of missingness in survey data is the _missing at random model_, which assumes that within classes the distribution of the data values that are missing for nonrespondents is the same as the distribution of data values for respondents. Thus, in replacing missing employment indicators, the researcher by using the survey data and his technical knowledge may decide that responses may be reasonably assumed to be missing at random within classes defined by age, race, and sex.

The third commonly used model for the nature of missingness in survey data is the regression model, which assumes that a linear model exists that can be used to predict the variable of interest based upon other items in the data record and that this model is the same for respondents and nonrespondents. For instance, the researcher might hypothesize that body weight can be predicted knowing age, sex, and body type and that the relationship between age, sex, body type, and weight should be similar for respondents and nonrespondents.

8.3 The No Imputation Procedure

The no imputation procedure refers to estimation where no attempt is made to replace missing data. No imputation is widely used in two basic variations. In the first variation of the no imputation procedure, nonrespondents are not included in the analysis and survey estimates are derived solely from respondent data. Thus, if medical expenses were being contrasted for males versus females, the researcher using this method would (1) eliminate those individuals with gender unknown or unknown medical expenses, and (2) summarize the medical experience of the remaining individuals for whom both gender and medical expense data were present. The second variation of the no imputation procedure creates "unknown" categories and reports statistics for these categories. Thus, "sex unknown" and "medical expenses unknown" categories would be included in comparing medical expenses by gender.

The effectiveness of the no imputation approach is dependent upon the parameters being estimated and the extent to which data are missing. The estimation of population totals is particularly sensitive to bias due to the presence of missing data. For instance, the number of males with high medical expenditures will be underestimated to the extent that sample males with high medical expenditures do not indicate that they are male or fail to provide medical expense data. For estimation of mean and proportions, the situation is somewhat different. In this situation, the estimated mean or proportion may be biased in either direction. To follow through with the previous example, the average medical expenses for males as estimated from the respondent group alone will be biased in an downward direction if nonresponding males tend to have larger medical expenses than responding males and biased upwards if the reverse is true. Only when nonresponding males have medical expenses comparable to those of responding males will the mean estimated from respondent data be unbiased with respect to inferences to the entire population of males. For the medical expenses example, this is unlikely to be the case since remembering expenditures may be more difficult when the amount is large than when the amount is zero.

The bias in no imputation survey estimates will depend upon the extent of missing data and the degree to which nonrespondents as a group differ from respondents as a group. As a rule of thumb, the authors suggest that the use of no imputation should probably be restricted to situations in which the level of missing data is small (less that five percent) and preferrably very small (less than two percent).

An example of when the no imputation approach is acceptable is furnished by an empirical investigation of item nonresponse to the National Longitudinal Survey of the High School Class of 1972 (Cox and Folsom, 1978.) When response rates were high, imputation often did not produce any gain in the accuracy of survey estimates. Cox and Folsom attribute the lack of important gains through imputation to the fact that reductions in bias were accompanied by offsetting increases in the variance of survey estimates. For income items where the level of missing data was larger, imputation improved the accuracy of survey estimates.

Even when the level of missing data is very small, imputation of missing data may be needed for pragmatic reasons. It is often desirable to have domain identifiers defined for the entire sample, for instance. When a post-stratification adjustment is made as a part of the weighting process to age, race, and sex population totals, all respondents must be assigned to an age, race, and sex cell in order to receive an analysis weight. Rather than discarding the entire data record for nonrespondents who fail to provide age, race, and sex data, the small percentage with these data missing can have values defined by imputation. Another reason to replace missing values is to avoid the confusion that results when estimates for population subgroups (males and females, for instance) do not add to the population total. This latter event can occur when individuals are included in the total population estimate but not in partitions due to missing data. Finally, complex analyses which use many variables, such as regression or factor analyses, may be easier to implement when missing values have been replaced. In addition, the effect of item nonresponse is likely to be cumulative so that analyses with several variables may have few data records with complete responses to all items, making some form of missing data imputation essential.

8.4 Logical Imputation of Missing Values

When data are missing for a particular item, there may be responses to other questions that allow the deduction of what the missing response should be. The process of using responses to related data items to logically determine the response that is missing is referred to as *logical imputation*. For instance, gender may be inferred from relationship to head (e.g., son, daughter) or from title (e.g., Mr., Mrs., Miss). Logical imputation is appropriate when a unique association exists between the predictor variable and the variable to be predicted. Thus, it would be reasonable to infer that sons and uncles were male or persons with titles of Ms., Mrs., or Miss were female. It is not so reasonable to impute the gender "male" to a respondent in the armed forces.

Even when good predictor variables are present, a prediction may not exist for every record with missing data. Some relationship to head responses (e.g., cousin) will not provide information as to gender, just as some titles (e.g., Dr.) provide no gender information. Thus, missing data will usually exist even after logical imputation has been completed. In this situation, the researcher must consider whether imputation is needed to replace missing responses.

8.5 The Mean Value Imputation Procedure

Perhaps the simplest imputation procedure to use is the mean value imputation procedure. The *mean value imputation procedure* replaces missing data with the average value of the respondent data. Thus, if the average of the employment incomes reported by survey respondents was $15,000 a year, then all income nonrespondents would have "$15,000" imputed as their income. A discrete analogue would be to impute the modal response. If female respondents were more common than male respondents, then gender nonrespondents would have "female" imputed as their gender. The imputation may be done within *imputation classes*, groups of people judged to have similar responses to the data item subject to imputation.

When mean value imputation is used for data items with continuous rather than discrete responses, the mean estimated from the imputation-revised data will be equal to the mean of the respondent data for each of the imputation classes. When imputation classes can be defined that are strongly related to the response being imputed, the mean estimated using the imputation-revised data should have much of the nonresponse bias removed.

When parameters other than means are being estimated, this method is likely to produce worse estimates than those derived from no imputation at all. For instance, if the distribution of income is being estimated, mean value imputation will result in overestimation of the percentage falling into the middle of the distribution and an underestimation of the percentage with high and low incomes. The deleterious effect of mean value imputation for distributional estimates severely limits its utility. Mean value imputation has been discussed in this chapter for the sake of completeness and is not recommended for general use.

8.6 The Cold Deck Imputation Procedure

The cold deck imputation procedure substitutes values from some previous census or survey for missing values for items on the current questionnaire. Using this procedure, a cold deck is formulated by classifying the previous data set according to the categories that can also be used to classify the present data set (e.g., race, sex, income, etc.). For each category that is defined, a distribution of responses is constructed based upon the older set of data. In processing the new data, when a missing response is determined for an item on a questionnaire, the imputation class to which the individual belongs is ascertained and a response is selected at random from the cold deck distribution for that imputation class. This response is then imputed (substituted) for the missing response in the current data set. After all questionnaires have been processed and a cold deck value imputed for each missing response, means and their variances are usually computed by ignoring the fact that an imputation procedure was used.

An advantage of the cold deck imputation method is that missing responses can be imputed as the data are being processed. But the procedure also increases the variance to an extent not reflected in the estimated variance. Other criticisms of the cold deck imputation technique include the procedure's heavy reliance on the accuracy and currency of the older set of data with respect to the new set of data and the fact that information from the current data are not used for imputation purposes. Not in common use today, this technique has been discussed to provide a historical background for the hot deck method.

8.7 The Hot Deck Imputation Procedure

The hot deck imputation procedure eliminates the criticism that the current data are not used for imputation purposes. This technique is similar to the cold deck procedure in that it allows imputation of missing responses as the data are processed. When timeliness is an important factor, such as in rapid turnaround surveys, this facet can become very important.

To implement the *hot deck imputation procedure*, the individuals completing the questionnaire are again partitioned into imputation classes. Often users sort the records within classes when there are other data available that relate to response. An initial value is determined for each class in the hot deck based upon previous data or the current data set. As new data are processed, the imputation class to which each individual belongs is determined. If the questionnaire being processed is complete, then that individual's responses replace the responses stored in the relevant class of the hot deck. Thus, new responses are supplied for each class of the hot deck as they appear in the data file. When a questionnaire is encountered with a missing item, the response in the same class of the hot deck is imputed for the missing response. When all questionnaires have been processed and the missing data imputed, means and their variances are usually computed without accounting for the effect of the imputation procedure.

The ease of use and flexibility of implementation of the hot deck technique has led to it becoming the most commonly used item nonresponse imputation procedure. However, Bailar and Bailar (1978) have demonstrated that the hot deck procedure will cause an increase in the variance of sample means compared to the no imputation procedure which ignores missing values in computing item means. Another flaw in the hot deck technique is that variance estimates cannot be obtained analytically but must be estimated using some form of pseudoreplication such as balanced repeated replication (BRR). In practice, most users of the hot deck ignore the fact that imputation has occurred and compute the variance in the usual manner. The resulting variance will typically underestimate the true variance of the sample statistic. Other flaws in the hot deck procedure are that there is no probability mechanism attached to the assignment of missing values and that the same individual's responses may be used repeatedly to supply missing information. Variations of the hot deck have been developed that retain more than one response per imputation class in order to reduce the use of multiple donors. These variations usually do not attach a probability mechanism to the assignment of missing donors.

8.8 The Weighted Hot Deck Imputation Procedure

When the level of missing data is large (ten percent or more), greater control is needed over the imputation process to lessen the variance inflation effect of imputation. In fact, the ideal estimation procedure may not include imputation at all. When individuals can be divided into classes within which respondents and nonrespondents will have similar responses to the data item being analyzed, a weight adjustment procedure will tend to produce estimates with minimum mean square error. A *weight adjustment procedure* would adjust the weights of item respondents within each weighting class by a common factor so that the adjusted weights for the item respondents sum to the population total for that class. These adjusted weights would then be used in analyzing the item data.

With this approach, the bias of nonresponse-adjusted estimates will depend upon the extent to which the nonrespondent distribution differs from the respondent distribution for each class. The weight adjustment will result in an increase in the variance of survey estimates due to unequal weighting but this variance inflation should be less than that resulting from most imputation procedures. Unfortunately, the weight adjustment approach is often infeasible for general survey use, since separate weights have to be developed for every combination of variables subject to analysis. Weight adjustments are also impractical when data recodes are to be created subsequently by aggregating across variables. In spite of the advantages of weight adjustment, most surveys need a general purpose imputation procedure that produces imputation-revised data that can be used across analyses.

The weighted sequential hot deck imputation procedure was designed to replace missing data while approximating the properties of a weight adjustment procedure (Cox, 1980). A sequential sample selection method developed by Chromy (1979) was adapted to create the procedure. Weighted hot deck imputation is designed so that means and proportions, estimated using the imputation-revised data, will be equal in expectation to the weighted mean or proportion estimated using respondent data only. Note that variances, co-variances, correlations, regression coefficients, and other higher order population parameters are estimated by simple functions involving weighted means of products and cross-products. Hence, the expected value of such higher order statistics for the imputation-revised data will also equal the corresponding estimator obtained from respondent data only. Achieving this desirable result implies that the analysis weights of respondents and nonrespondents have to be simultaneously considered so as not to seriously bias the estimated distribution of variates obtained from respondent data.

An objection to the usual unweighted procedure is that the data from a respondent can be used many times for imputation since the data from a respondent record will be imputed to every nonrespondent record following in the data file, until another respondent record is encountered. The weighted sequential hot deck procedure has the additional advantage that it controls the number of times that respondent records are used for imputation.

The imputation strategy may be thought of as utilizing two data files: a data file of respondents and a data file on nonrespondents. The imputation procedure substitutes data values of responding individuals for the responses that are missing for nonresponding individuals. The first step in the process is to sort the two data files with respect to variables related to response and the distribution of characteristics of interest. Both files will have analysis weights attached to each individual. (For a census or a self-weighting sample, weights equal to one may be used.) The imputation occurs within imputation classes so that the distribution of means and proportions is preserved within each class over repeated imputations.

The implementation of the weighted sequential hot deck imputation technique within classes proceeds as follows. In sequential order, the data associated with respondents are assigned to nonrespondents. The number of times that the data record for respondent i is accessed to impute to nonresponding individuals, or n(i), is defined as a function of the analysis weight of respondent i and the analysis weights of the nonrespondents to which it can potentially be linked for imputation.

Having determined the number of times each respondent's record is to be used for imputation, missing data are imputed as follows. The first n(1) nonrespondents have their missing data replaced by data imputed from the first respondent. The next n(2) nonrespondents have data imputed from the second respondent and so on until the last n(r) nonrespondents have their missing data replaced by data imputed from the r-th respondent (the last respondent) in the respondent data file.

The weighted hot deck approach is not difficult to implement. Using the computer software for Chromy's sequential sample selection algorithm as a starting point, Iannacchione (1982) developed general purpose software that performs weighted sequential hot deck imputation. This software has been implemented for data items from several studies and the results evaluated (e.g., Cox and Folsom, 1981 and Williams and Folsom, 1981). Programming for weighted imputation is not significantly more time consuming than programming a standard hot deck imputation. In addition, the weighted approach provides control over the imputation process that can become quite valuable when response rates are low.

8.9 Regression Imputation

Regression imputation is most appropriate for data items with quantitative responses when there are other quantitative variables that can be used to predict missing responses. Suppose that preliminary research indicates that a fairly strong linear relationship exists between the response variable y and another continuous variable x. This relationship is observed for the respondent data only but the researcher hypothesizes that this relationship also exists for the data from nonrespondents. In this situation, one can use the respondent data to fit the following model:

$$\underset{\sim}{y} = \alpha + \beta\underset{\sim}{x} + \underset{\sim}{\varepsilon} \tag{8-5}$$

where α and β are the intercept and slope for the regression line.

Missing responses for individual i can then be replaced by the predicted response or

$$\hat{y}_i = \alpha + \beta x_i \tag{8-6}$$

The regression approach to imputation should be considered when there are data items available for nonrespondents as well as respondents that the researcher believes can be used to model the response variate for which data are missing. The independent variables used in the regression may have qualitative or quantitative responses. Qualitative variables are incorporated into the regression equation by creating zero-one indicator variables for each level of the qualitative variable.

The construction and evaluation of a good regression model is time consuming but has a greater potential of producing imputed values that are closer to the true value than other direct imputation approaches. However, the researcher should be aware that he is actually imputing a mean value when using regression imputation. If income is predicted based upon age, race, and sex, for instance, every individual with missing data of a certain age, race, and sex will have the same value imputed. When the analysis plan involves estimation of distributional statistics, regression imputation of missing values can distort the distribution and hence result in biased survey estimates. In this situation, the researcher can consider modifying the regression-based imputed data by including an estimated residual. Perhaps the best approach would be to use some form of hot deck imputation to associate residuals for respondents to nonrespondents. When weighted sequential hot deck imputation of residuals is used, the distribution of the respondent data will be preserved.

8.10 Concluding Remarks

Survey data are prone to error from a number of sources. Respondents may misunderstand the question being asked. The interviewer may incorrectly record the respondent's answer or the keyer may miskey the response. All of these error sources result in erroneous data in the data base. Similar errors can also produce missing data. That is, the respondent may not answer a question, the interviewer may forget to ask a question or may neglect to record the response, or the keyer may overlook the interviewer-recorded response.

The first step in dealing with item nonresponse and inconsistencies in the responses to data items should be to check the questionnaire to determine if logical, deductive imputations can be made for the missing or inconsistent data based upon the responses to other questions. If logical imputations cannot be made, the researcher utilizing the data should consider whether imputation is needed to replace missing or inconsistent responses.

There are a plethora of imputation procedures which may be used to replace missing item data. Except for the cold deck and mean value imputation which were described for historical background, one of the imputation procedures discussed in this report should be suitable for most survey analysis needs. The unweighted hot deck approach is easiest to use both in terms of programmer and computer time needed to implement the procedure. It is most appropriate when dealing with qualitative variables and when the level of missing data is low. The weighted hot deck procedure is a little more difficult to implement but the greater measure of control it provides makes it ideal when response rates are low and good predictive variables are not available for model building. The regression procedure is useful when dealing with quantitative variables for which a good predictive regression equation can be developed. It can be the most time consuming of the three approaches and hence should probably be reserved for data items for which greater precision is necessary.

The literature provides little guidance on when to impute and when not to impute. This is mainly due to the fact that nonresponse produces a Catch 22 situation in which the impact of nonresponse could be estimated if the missing data were known but if the missing data were known there would be no nonresponse. Some authors have tried to overcome this problem by simulation of the effect of nonresponse (e.g., Ford, 1976 and Kalton, 1983). These studies provide insight into the effect of missing data but cannot totally reflect the complex nature of missing data. Cox and Folsom (1978) present the results of a study of alternative item nonresponse adjustment procedures for survey data in which missing and inconsistent data were obtained by telephone follow-up. In this case the true bias of the survey estimates could be obtained. Based upon the results of these three studies, it would appear that imputation should be considered when survey estimates would be significantly biased without such corrections. If nonresponse bias is small, imputation can increase the variation of survey estimates and produce estimates with a greater mean square error. Additional information about imputation methods is provided by Sande (1982) and by the National Research Council's Panel on Incomplete Data (1983) in Volume 2 of their three volume set, Incomplete Data in Sample Surveys.

References

Bailar, J. C. and B. A. Bailar (1978). Comparison of Two Procedures for Imputing Missing Survey Values. Proceedings of the American Statistical Association, Survey Research Methods Section, 462-467.

Chapman, D. W. (1976). A Survey of Nonresponse Imputation Procedures. Proceedings of the American Statistical Association, Social Statistics Section, 245-249.

Chromy, J. R. (1979). Sequential Sample Selection Methods. Proceedings of the American Statistical Association, Survey Research Methods Section, 401-406.

Cox, B. G. (1980). The Weighted Sequential Hot Deck Imputation Procedure. Proceedings of the American Statistical Association, Survey Research Methods Section, 721-726.

Cox, B. G. and R. E. Folsom (1978). An Empirical Investigation of Alternate Item Nonresponse Adjustments. Proceedings of the American Statistical Association, Survey Research Methods Section, 219-223.

Cox, B. G. and R. E. Folsom (1981). An Evaluation of Weighted Hot Deck Imputation for Unreported Health Care Visits. Proceedings of the American Statistical Association, Survey Research Methods Section, 412-417.

Ford, B. (1976). Missing Data Procedures: A Comparative Study. Proceedings of the American Statistical Association, Social Statistics Section, 324-329.

Iannacchione, V. G. (1982). Weighted Sequential Hot Deck Imputation Macros. Proceedings of the Seventh Annual SAS User's Group International Conference, 759-763.

Kalton, G. (1983). _Compensating for Missing Survey Data_. Research Report Series, Institute for Social Research, The University of Michigan, Ann Arbor, Michigan.

Panel on Incomplete Data, National Research Council (1983). _Incomplete Data in Sample Surveys_. W. G. Madow, I. Olkin, and D. B. Rubin editors, New York: Academic Press.

Sande, I. G. (1982). Imputation in Surveys: Coping with Reality. _The American Statistician_, 36, 145-152.

Williams, R. L. and R. E. Folsom (1981). Weighted Hot Deck Imputation of Medical Expenditures Based on a Record Check Subsample. _Proceedings of the American Statistical Association, Survey Research Methods Section_, 406-411.

9
Imputation Illustrated for Demographic Items

Almost all researchers involved in the analysis of data from sample surveys have faced the problem of the proper approach to use in analyzing data sets that contain missing and obviously erroneous responses. Response errors can be minimized through effective instrument and interview design. However, even the best designed surveys will suffer from some nonresponse, especially for quantitative or sensitive items. Associated with the problem of item nonresponse is that of inconsistent or invalid responses and violations of skip patterns. These response errors are an obvious source of bias which must be considered in data analysis.

To assist data analysis, the NMCUES developed a public use file for which the most important variables had (1) data inconsistencies and invalid data removed through logical editing, and (2) missing data replaced using one or more of the imputation procedures described in the previous chapter. One important result of this process was the recognition that the easiest and least costly step in creating analysis variables is the actual imputation of missing data. Rather, it is the steps leading to the construction of the input variable to the imputation process - the cleaned, edited variable with missing data indicated - that tend to be both complicated and costly in terms of person and computer time. In addition, the questionnaire must be carefully constructed to insure that required data are obtained so that the accuracy of analysis variables can be properly evaluated and missing data predicted.

The complexity involved in obtaining accurate data for analysis merits a discussion of the steps from design of the questionnaire data items through the editing and imputation needed to construct the final analysis recode variables. This chapter describes procedures that were used to replace missing data for specific questions in the NMCUES (Cox, et al, 1982; Cox, 1983). Data items have been chosen that illustrate the variety of situations that lead to missing data and how the nature of the missing data dictates the selection of the appropriate imputation procedure. The chapter closes with a discussion of the implications of using imputation-revised variables in statistical analyses.

Particular attention is placed on providing information for the following questions:

- What are the sources of faulty data and how do faulty data relate to the manner in which the data were collected?
- What logical editing operations can be used to improve the quality of the data?
- What is the nature and extent of the nonresponse observed?
- What variables are useful in predicting missing responses?

Since all of these questions are specific to the type of data, these topics are addressed by using the individual NMCUES data types as examples. Many of these data types are common across sample surveys (e.g., demographic variables); other data types are relevant for health surveys only. However, the principles followed in producing analysis variables are generally applicable to the problems that a researcher must address in producing imputation-revised variables for use in analysis.

9.1 Background of NMCUES Item Imputation

Health care estimates obtained using household data are known to be subject to bias due to the inability of many respondents to accurately recall past health care events. Household-reported expenditure and utilization data are especially prone to response errors since individuals may not be able to recall the charges associated with visits or they may fail to recall all visits that were made or may report visits that were not made during the reference period. To minimize such sources of response error, NMCUES included two types of memory aids. First, during the Round 1 interview each household was given a calendar/diary which included a calendar for each survey month with space for noting dates of illnesses and medical provider visits, a monthly diary section for noting the costs of medical provider visits and prescribed medications, and a pocket for storing medical bills and receipts. Second, computer-generated summaries of previously reported medical care visits and expenditures were sent to each household and to the interviewer prior to the start of each data collection round. The summaries gave the respondents a chance to review data reported in previous interviews and to add, delete, or change incorrect or incomplete data on medical care utilization and expenditures. The final updated summary was used to construct the summary portion of the medical care visit-level files. In addition to these memory aids, NMCUES also used a relatively short recall period of twelve weeks to facilitate recall of medical events.

However, the respondents could not report data that were unknown to them. In other instances, the respondents did not know (or chose not to state) the charges associated with a visit. Items that were derived by summing data over data collection rounds suffered from nonresponse due to survey attrition. In addition, the sensitive nature of some of the data being requested (e.g., income) resulted in higher levels of nonresponse for these items.

Because of the size and complexity of the NMCUES data base, it was not feasible, from a cost standpoint, to replace all missing data for all data items. Neither was it reasonable to use only those records with absolutely complete data; over the 1,400 NMCUES data items only a few individuals could be expected to provide complete data. With these facts in mind, the NMCUES approach was to designate a subsample of the total items on the data base as important enough to merit missing data imputations. For five percent of the NMCUES data items, the responses were edited and missing data replaced in order to produce imputation-revised variables for use in analysis. Items for which imputations were made cover the following data areas:

- age and birthdate,
- race and Hispanic origin,
- sex,
- educational level,
- employment status,
- disability days,
- nights hospitalized,
- health insurance,
- income, and
- health care charges.

Because the number of items subject to imputation was large (320), the items were divided into sets and imputations performed within these sets. Every imputation and editing operation was conducted independently within the samples for the National Household Survey (HHS) and the State Medicaid Household Surveys (SMHS). For each sample, the data records were usually partitioned into classes related to response with the imputation occurring independently within each class. The data records were also sorted within each class for additional control over the imputation process.

For each group of data items, the most cost effective imputation strategy was selected. The large number of items for which imputations were required and the costs associated with processing the large files necessitated that alternative approaches be evaluated in terms of the quality of the imputations and the associated costs. By reducing the number of passes through the data, some approaches could drastically reduce data processing costs while producing results that were essentially comparable in quality to the best approach. The remainder of this chapter describes the specific approach used for selected data types with an emphasis on (1) methods to collect the data, (2) sources of missing and faulty data, (3) cleaning and editing needed to define analysis recodes, and (4) the set up of the imputation procedure.

9.2 Age and Birthdate Questions

Age-defined reporting domains are among the more important domains for NMCUES analyses. The NMCUES included questions that asked the age of each sample individual as well as his birthdate. Both questions were asked because of a perception that individuals would report birthdate more accurately than age but when proxy data were collected for an individual only age might be known. Since many NMCUES skip patterns were based upon age, the age variable was updated in each data collection round. Hence, the survey-reported age was not age as of a particular date but age of the last interview. For analysis purposes, NMCUES needed a revised age variable; this age recode was defined as the age of the individual on January 1, 1980. Birthdate was used to define the revised age variable with 1980 newborns assigned a special consistency code to indicate that they were born after January 1. When birthdate was missing, the survey-reported age was used to define the recode.

Prior to replacing data missing for the age recode, the recode was checked to determine if logical editing or imputation was needed by comparing the survey-reported age to the age recode. Out of

32,874 individuals in the combined HHS and SMHS samples, 318 had the birthdate-oriented age recode disagreeing with the survey-reported age by more than three years. To determine whether to believe the age or the birthdate, the data records for these individuals were printed with the age recode, the survey-reported age, the birthdate, and the individual's relationship to the head of household.

The most common discrepancy had the birthdate as 1980, which indicated that the individual was a newborn, but did not have associated with their data the special code that NMCUES used for newborns. Many of these individuals had as their relationship to household head either head, father, mother, grandfather, or grandmother. With the relationship to head agreeing with the survey-reported age and no other indication that they were newborns, it appeared that the interviewers had made a transcription error and accidentally recorded the current year (1980 when data collection occurred) rather than the year of birth. For 49 such records, the birthyear and the age recode were changed so that they agreed with the survey-reported age.

The second most common discrepancy had the age recode differing from the survey-reported age by a factor of 10 or 100. For instance, the year of birth on one record was given as 1875, resulting in an age recode of 105, and the age as of the last interview was given as six, indicating that the individual was probably 5 on January 1, 1980. To resolve these kinds of inconsistencies, the age recode was compared to the relationship to head to see if the birthyear and age recode should be revised. A total of 14 records were revised as a result of this process. For the previous example, the relationship to head was "son" so the birthyear was changed to 1975 and the age recode to 5.

After this logical editing and imputation had been completed, the age recode variable was defined for over 99 percent of the HHS and SMHS sample individuals. With only 60 individuals with missing data for the age recode, nonresponse bias would be negligible except when age-related domain totals were being estimated. However, since the sample weight development included a post-stratification adjustment to age category totals, an age recode value was needed for all individuals to assign them to post-strata.

An unweighted hot deck imputation procedure was appropriate to impute an age to these individuals since the level of missing data was low. For better control over the imputations, the data records were sorted by relationship to household head and by age of the head prior to the imputation. Hence, the value imputed to a sample individual was taken from someone adjacent in the data file, which implies that the donor of the age recode value had a similar relationship to their household head and a similar age for the head. As with all NMCUES imputations, the imputation was performed separately for the five samples (the HHS and four SMHS samples), since each sample was based upon a different target population. Table 9-1 summarizes the results of the imputation procedure.

When discrepancies between the birthdate-derived age recode and the survey-reported age led to a revision of the age recode, the year of birth was revised as well. After imputation was completed

Table 9-1. Imputation Results for Selected NMCUES Questions

Item/Imputation Result	Sample Type				
	HHS	SMHS CA	SMHS MI	SMHS NY	SMHS TX
Age:					
Real Data	99.8	99.3	99.8	99.5	99.8
Logical Imputation	0.1	0.1	0.1	0.2	0.1
Statistical Imputation	0.1	0.6	0.1	0.3	0.1
Birthdate:					
No Field Imputed	99.1	98.2	98.9	98.8	98.9
Some Fields Imputed	0.9	1.8	1.1	1.2	1.1
Race:					
Real Data	80.1	81.6	77.0	74.6	68.4
Logical Imputation	19.7	17.6	22.9	23.1	31.1
Statistical Imputation	0.2	0.8	0.1	2.3	0.5
Hispanic Origin:					
Real Data	79.9	80.0	76.5	74.6	68.4
Logical Imputation	19.8	18.3	23.0	23.0	31.1
Statistical Imputation	0.3	1.7	0.5	2.4	0.5
Sex:					
Real Data	98.3	99.3	99.5	99.3	99.5
Logical Imputation	1.2	0.3	0.2	0.2	0.3
Statistical Imputation	0.5	0.4	0.3	0.5	0.2
Education:					
Real Data	99.0	97.0	97.7	97.6	98.6
Logical Imputation	0.1	0.1	0.1	0.1	0.0
Statistical Imputation	0.9	2.9	2.2	2.3	1.4

for the age recode, the age recode was used to define year of birth for those individuals with missing data for the year. Any records with missing month or day fields had the data replaced using an unweighted hot deck procedure. Referring to Table 9-1 again, at least one of the month, day, or year fields was imputed for 0.9 percent of the HHS sample and for 1.1 to 1.8 percent of the SMHS samples. After all data processing was completed, the imputations were checked by comparing the age recode to the age as defined by the imputation-revised birthdate.

9.3 Race and Hispanic Origin Questions

Obtaining racial classification presents definitional as well as operational problems. With respect to definition, a decision needs to be made as to the proper classification of individuals of mixed origin. Operationally, the interviewer can be asked to identify the respondent's race or the respondents may be asked to classify themselves. Classifying children of racially mixed marriages as to race can be more difficult and, unless tact is used, the parents may become defensive.

To solve these problems, NMCUES adopted the following data collection strategy. During the first interview, the respondent was handed a card and instructed, "Please look at this card and tell me the number of the group or groups which describes (PERSON'S) racial background." The following categories were given: (1) American Indian or Alaskan Native, (2) Asian or Pacific Islander, (3) Black, (4) White, and (5) Other (SPECIFY). If the respondent mentioned more than one code, the interviewer then asked which one best described the respondent's racial background. Previous to data collection, NMCUES had decided that the race of children would not be asked. Instead, the race of a child would be defined to be that of the mother or of other household members. However, the interviewer was instructed to observe and record each individual's race on the control card with the following categories: white, black, or other.

These data were used to construct a race recode variable. The variable was defined using the respondent-reported racial classification. If more than one race were indicated and the respondent did not choose one that best described him, the race recode was defined using the priority ordering: (1) Black, (2) American Indian or Alaskan Native, (3) Asian or Pacific Islander, and (4) White. It is important to note that these four race codes form the totality of possible race groups. For this reason, the "Other (SPECIFY)" responses were examined by data editors and reclassified into one of these four racial groups prior to defining the race recode. The "other" responses that could not be reclassified were treated as missing. In general, the "other" responses were the result of an individual reporting his nationality (e.g., Albanian) rather than his race.

At this point, all children under 17, as well as a small proportion of those 17 and older, had missing data for the race recode. Although the interviewer-observed race was available, it was decided not to use these data to logically create the race recode for two reasons. First, the interviewer assigned race as Black, White, or

Other. Hence, the "Other" category was a combination of American Indian or Alaskan native, Asian or Pacific Islander, and unclassifiable individuals. More importantly, the decision was made not to confuse the race recode definition by combining respondent-reported and interviewer-observed data.

Instead of directly imputing for a missing race recode based upon interviewer-observed race for the individual, the race recode was imputed from that of another household member. To impute the race, the race recodes of household members were examined in the following order until a nonmissing response was encountered: wife of head, head, and all other individuals. This first encountered value was used to impute for all missing race recode values within the household. Table 9-1 summarizes the results of this logical imputation. Race recode data were available for 80.1 percent of the HHS sample and for 68.4 to 81.6 percent of the SMHS samples. In the HHS, 19.7 percent had their race defined based upon that of a household member and 17.6 to 31.1 percent of the SMHS samples. The remaining 0.2 percent of the national sample and 0.1 to 2.3 percent of the SMHS samples had their race imputed using unweighted hot deck imputation. The imputation was done separately within each of the five sample types after the records were sorted by the interviewer-observed race variable for the household head, and then by household and geographical region. This procedure insured that the same race was imputed to all household members with missing data.

As a check of the quality of the imputations, the race recode variable was compared to the interviewer-observed race variable, before and after imputation of missing race recodes. Table 9-2 presents a comparison of the interviewer-observed race to the respondent-reported race (the race recode prior to imputation) for each of the five NMCUES samples. The entries in the table are the percent of the sample records falling into each cell. Only sample records with both variables defined were used in constructing Table 9-2. Note that the interviewer and the respondent agreed about race 97.8 percent of the time for the national sample and 91.8 to 98.3 percent of the time for the SMHS samples.

As indicated previously, the respondent was not asked to report race for children, which resulted in a large rate of missing data for respondent-reported race. Instead, the race for children and all other individuals with missing data was imputed based upon the race of other household members, whenever possible. To verify that this was a reasonable strategy, the distribution of the interviewer-observed race and the pre-imputation race recode was compared with the distribution when the imputation-revised race is used. Table 9-3 presents the cross-classification of the interviewer-observed race to the imputation-revised rates. The entries in the table are the percent of the total sample falling in the cell. Only records with the interviewer-observed race recode defined were used to construct the table. The interviewer-observed race was defined for 98 percent of the national sample and from 96 to 99 percent for the SMHS samples. Comparing Table 9-3 with Table 9-2, one can see that the distribution of the individuals across the cells defined by the cross-tabulation of race recode and interviewer-observed race is very similar before and after imputation revision of the race re-

Table 9-2. Distribution of Interviewer-Observed Race With Respect to Respondent-Reported Race

Respondent-Reported Race	Interviewer-Observed Race			Total
	White	Black	Other	
National Sample:				
White	86.1	0.1	1.0	87.2
Black	0.1	10.4	0.0	10.5
Asian or Pacific	0.3	0.0	1.2	1.5
Indian or Eskimo	0.5	0.1	0.1	0.7
Total	87.0	10.6	2.4	100.0
SMHS California:				
White	65.8	0.1	6.3	72.1
Black	0.1	19.4	0.0	19.5
Asian or Pacific	0.6	0.1	6.0	6.8
Indian or Eskimo	0.9	0.0	0.6	1.6
Total	67.5	19.7	12.9	100.0
SMHS Michigan:				
White	57.1	0.3	0.9	58.3
Black	0.1	39.2	0.0	39.3
Asian or Pacific	0.1	0.1	0.1	0.3
Indian or Eskimo	1.4	0.5	0.2	2.1
Total	58.7	40.1	1.2	100.0
SMHS New York:				
White	59.8	1.4	3.4	64.6
Black	0.5	30.8	0.3	31.6
Asian or Pacific	0.5	0.0	2.1	2.6
Indian or Eskimo	0.5	0.5	0.3	1.3
Total	61.3	32.6	6.1	100.0
SMHS Texas:				
White	65.7	0.2	0.6	66.5
Black	0.1	32.2	0.0	32.3
Asian or Pacific	0.2	0.0	0.4	0.6
Indian or Eskimo	0.5	0.1	0.0	0.6
Total	66.4	32.6	0.9	100.0

Note: Detail may not add to total shown because of rounding.

Table 9-3. Distribution of Interviewer-Observed Race With Respect to Imputation-Revised Race

Respondent-Reported Race	Interviewer-Observed Race			Total
	White	Black	Other	
National Sample:				
White	84.5	0.2	1.3	86.1
Black	0.2	11.5	0.0	11.7
Asian or Pacific	0.3	0.0	1.2	1.5
Indian or Eskimo	0.6	0.0	0.1	0.8
Total	85.6	11.8	2.6	100.0
SMHS California:				
White	64.1	0.2	7.9	72.2
Black	0.3	18.8	0.0	19.0
Asian or Pacific	0.6	0.2	5.9	6.7
Indian or Eskimo	1.5	0.0	0.6	2.1
Total	66.4	19.1	14.4	100.0
SMHS Michigan:				
White	56.2	0.4	1.0	57.6
Black	0.2	39.7	0.0	39.9
Asian or Pacific	0.3	0.2	0.1	0.6
Indian or Eskimo	1.2	0.5	0.2	1.9
Total	57.9	40.8	1.3	100.0
SMHS New York:				
White	57.7	2.0	4.2	63.9
Black	0.4	31.4	0.3	32.1
Asian or Pacific	0.5	0.0	1.9	2.4
Indian or Eskimo	0.6	0.4	0.6	1.7
Total	59.1	33.8	7.1	100.0
SMHS Texas:				
White	63.4	0.3	0.6	64.4
Black	0.1	34.1	0.0	34.2
Asian or Pacific	0.3	0.0	0.3	0.6
Indian or Eskimo	0.7	0.1	0.0	0.8
Total	64.5	34.6	0.9	100.0

Note: Detail may not add to total shown because of rounding.

code. The percent agreement between the race recode and the interviewer observed race was also preserved. After imputation, 97.3 percent of the national sample had the codes agreeing and from 89.4 to 97.8 percent of the codes agreed for the SMHS samples.

The Hispanic Origin of each person was determined by handing the respondent a card and asking, "Are any of these groups (PERSON'S) main national origin or ancestry?" The card listed the following groups: (1) Puerto Rican, (2) Cuban, (3) Mexican, (4) Mexicano, (5) Mexican-American, (6) Chicano, (7) Other Latin American, and (8) Other Spanish. Those who responded "Yes" were asked to give the number of the group to which they belonged.

The Hispanic Origin questions were phrased based upon past studies of the way individuals respond. For more convenient use in analysis, a Hispanic Origin recode was created with the following levels: (1) Non-Hispanic, (2) Puerto Rican, (3) Cuban, (4) Mexican or Mexican-American, and (5) Other Hispanic. The Non-Hispanic category was based upon a "No" response to the Hispanic Origin question with one exception. Occasionally an individual would have a particular Hispanic group marked without a "Yes" response to the Hispanic Origin question. This discrepancy was regarded to be the result of keying or transcription errors in the Hispanic Origin question since the interviewer should not have asked for a nationality group unless the individual indicated he was of Hispanic Origin. A similar strategy was used in developing other recode variables when skip pattern discrepancies were involved.

Again, the interviewer was instructed to skip this group of questions for those individuals less than age 17 who were not household heads. The Hispanic Origin recode for these individuals and for any others who did not respond was first determined from other members of the individual's household. An unweighted hot deck procedure was appropriate to impute for any remaining missing data since the missing data occurred at a low level. Classes were formed using the interviewer-observed race variable from the control card for the head of the individual's first encountered household. The file was sorted by household and geographical variables prior to doing the imputation.

These procedures implied that the same Hispanic Origin would be imputed to all household members with missing data and that the value imputed would be from someone of the same race who resided nearby. As seen in Table 9-1, Hispanic Origin data were available for 79.9 percent of the HHS sample and 68.4 to 80.0 percent of the SMHS samples; another 19.8 percent of the HHS sample had their Hispanic Origin assigned from other members of their household and from 18.3 to 31.1 percent of the SMHS samples. Hot deck imputation was needed for only 0.3 percent of the national sample and for 0.5 to 2.4 percent of the SMHS samples.

9.4 Sex Classification Questions

The classification of each sample individual as to male or female was done by the NMCUES interviewer. Because of a perception that it might be offensive to some individuals to ask them their sex, the interviewer was instructed to determine sex based upon physical appearance, relationship to head, name, and pronouns used to refer to the person and to ask the sex if it could not be ascertained otherwise. Although some might argue that asking a person's sex is not offensive when it is done in a routine, filling-out-forms manner, the NMCUES approach should also be satisfactory since there were five rounds of data collection so that errors could be corrected later.

In replacing missing data for the sex variable, the relationship to household head was examined for all data collection rounds to see if any relationships provided information as to sex (e.g., wife, son, father, etc.). The first encountered sex-defining relationship was used to replace missing data for the person's sex. As seen in Table 9-1, the sex variable was defined for 98.3 percent of the HHS sample; an additional 1.2 percent had their sex defined based upon their relationship to head. For the remaining 0.5 percent of the HHS sample, an unweighted hot deck was again used to replace the missing values since the level of missing data was low. Over the four SMHS samples, sex data were obtained for 99.3 to 99.5 percent of the individuals with an additional 0.2 to 0.3 percent with sex defined based upon relationship to household head.

9.5 Educational Level Questions

For NMCUES analyses, the highest grade completed in school was needed for each individual 17 years old or older and for all heads of households. The required data was obtained by asking the "highest grade attended in school" and "did (PERSON) complete the highest grade attended?" The responses to these items were combined to create a highest grade completed in school recode. For minors who were not household heads, the recode was set to a consistency code to indicate that the recode was not applicable for them. At this point, logical imputation was used to assign the highest grade completed for 34 persons who were known to be college students. Age was used to decide the year of college to impute. For the remaining individuals, a value was assigned using an unweighted hot deck procedure within classes defined by race and sex; within each class the records were sorted by age. The imputed-revised data were available at this point so that missing data for classing and sorting variables was not a problem. Of the HHS sample, 99.0 percent provided data or were minors for whom the question was not applicable. For the remaining 1.0 percent, a value was imputed through logical or hot deck imputation. As seen in Table 9-1, from 97.0 to 98.6 percent of the SMHS samples provided education data.

9.6 Income Questions

Income data have traditionally been subject to higher levels of nonresponse. Not only is income a sensitive subject to discuss in a sample survey but income is also something not readily known. Interest income is difficult for most respondents to report since it varies over the year and many individuals do not take notice of the exact amount they receive. Employment income is usually better known but still presents problems. For instance, an individual may know his hourly wage but may not be aware of how many hours he worked during the year. If a raise was received during the year, arithmetic may be needed to determine the employment income for the year.

NMCUES used several approaches to insure that 1980 income data were as complete and accurate as possible. Income data were collected in Round 5 of the survey. At the time Round 5 data collection began in 1981, most individuals should have received their 1980 employment income statements (W2 Forms) and 1980 interest payment records from their banks and savings institutions. These summaries were available for the respondent to check. Even those respondents who replied without consulting their records could be expected to have reviewed the forms as they were received so that they should have a basis for reporting these two sources of income.

In addition to gathering income data when tax records were available, NMCUES also gathered ancillary data that could be used to estimate or impute income when data were not provided. Wage and salary information was requested in Round 5 for main and second jobs. In each round of the survey, the weeks and hours per week worked were requested for the main and second job. In Round 1 of the survey, the individual's usual activity in 1979 was obtained. This variable had levels denoting housewife, retired, student, employed, and something else. The total 1979 income was also requested. These data proved quite useful in the creation of imputation-revised income data.

NMCUES gathered 1980 income data for twelve income sources: money received from working, veteran's payments, unemployment compensation, worker's compensation, Supplemental Security Income,

Social Security Income, public assistance income, pension income, cash payments such as alimony and child support, interest income, investment and rental income, and all other income. Several sources of missing data existed for these variables. Since the data were obtained in Round 5, data were automatically missing for individuals who dropped out of the survey early or who became ineligible through death, institutionalization, or nonresidence in the United States. For reasons such as these, approximately five to ten percent of the samples had no income data at all. For the remaining individuals, certain income items had missing data and others had responses.

Since the questions were phrased to determine the presence or absence of a source of income first and then the amount if income was indicated, two types of missing data existed. One type of missing data had the individuals indicating that they received income from a source but the amount was not provided. The second type had the amount missing as well as missing data for the indicator of presence or absence of income. The first step in the income imputation was to create two recodes for each income source, one that gave income amount and another that specified whether or not income was received from that source. Data could be missing for one or both of these recodes.

In constructing the employment income recode, the response to a direct question asking employment income was used as long as valid data were obtained for the question. When no valid data were available for the employment income amount question, the employment income was set to zero when the total hours worked in 1980 was zero or the individual was less than 14 years old. When the total hours worked in 1980 were greater than zero and hourly pay was given, the 1980 employment income was calculated as the product of the two. For the remaining individuals who had hours worked greater than zero, the employment income amount was set to missing. Due to the use of the imputation-revised employment data to determine when an individual worked (and hence received employment income), there was no need to create a special income indicator variable since all individuals could be classified as to whether or not they had employment income.

For the remaining income sources, the presence or absence of income was determined using two variables. One variable was for the household and recorded whether or not anyone within the household had income from the source. For households that reported this income, the status of each individual within the household with respect to receiving income from the source was determined. Those individuals with a "Yes" response to the last question were also given a positive response for the income indicator. Those individuals who had a "No" response to either question were given a negative response to the income indicator.

At this point, the income amount was defined for each source based upon the income amount question(s). Those individuals with "No" for their income indicator were assigned zero as their income. In the ambiguous instance in which the response was "No" for the income source and an income amount was provided, the latter positive answer was given precedence. This is another example of the use of the skip pattern rule in imputation, which uses the response within a skip pattern before the lead in question when the two are in disagreement.

In some instances, the income amount had to be calculated. For the Supplemental Security Income, for instance, where the amount can vary, respondents were allowed to report four different payment amounts for four different time periods. This approach was used to allow for responses such as, "I received $100 a month until May and thereafter I received $150 a month until last December when I went down to $125 a month." This approach can result in unidentifiable cases of missing data in the sense that when three time periods and amounts are given, for instance, there is no way of telling whether the fourth time period and amount are missing or that there was no fourth time period. The NMCUES approach was to treat the income data as totally missing when there was a nonmissing time period with a missing amount. Otherwise, the income data were calculated as the sum of the products of the nonmissing time periods and the associated amounts.

At this point, missing data occurred for all of the income recodes and all of the income indicators except for employment. The imputation of missing data was done in several steps because of the varying amount of auxiliary data available for some income types and the varying degree to which nonresponse was a problem. It was recognized that the amount received for the various income items tend to be related. For instance, persons receiving public assistance or Social Security income receive little if any income from employment, and vice versa. Also persons with large employment incomes also tend to have larger amounts of interest and dividend income. Because of these relationships, the decision was made to replace missing data for employment income first and then use employment income in predicting the remaining income items.

Missing employment income was replaced using a weighted sequential hot deck approach due to the need for control resulting from the larger level of missing data. For the national sample, classes were defined based upon age, sex, occupation, and a categorized version of hours worked. Within each class, the records were sorted by years of education (0-8, 9-11, 12, 13+) and by the total hours worked in 1980 (uncategorized). For the State Medicaid Household Surveys, occupation and education were not used; otherwise, the classing and sorting variables were the same. Education and occupation were not used in classing due to the more homogeneous nature of the Medicaid population with respect to these variables. Employment income was imputed for 10.2 percent of the national sample and from 6.8 to 9.5 percent of the State Medicaid samples (Table 9-4). The rate of missing data for the State Medicaid samples can be expected to be lower since more individuals report no income.

For the remaining income items, two different types of missing data occurred. Some individuals dropped out of the survey prior to Round 5 and hence no income data were obtained. Other individuals completed the Round 5 Supplement but were unable or unwilling to answer all of the income items. The general approach for imputation was to use employment and other auxilary data to replace the missing values via a weighted sequential hot deck imputation procedure.

Table 9-4. Response Rates to Income Items for the NMCUES*

Income Type and Sample	Income Amount		
	Zero	Nonzero	Total
Employment			
HHS	99.8	80.7	89.8
SMHS CA	100.0	69.8	90.8
SMHS MI	100.0	65.6	90.5
SMHS NY	100.0	68.4	93.2
SMHS TX	99.9	67.1	91.4
Veteran's Pay			
HHS	94.6	87.5	94.4
SMHS CA	93.2	86.9	93.1
SMHS MI	93.6	87.5	93.5
SMHS NY	95.6	89.7	95.5
SMHS TX	94.9	80.0	94.8
Unemployment Benefits			
HHS	94.6	92.4	94.5
SMHS CA	93.2	85.7	93.0
SMHS MI	93.6	89.8	93.4
SMHS NY	95.6	95.3	95.6
SMHS TX	94.9	93.8	94.9
Workman's Compensation			
HHS	94.6	92.4	94.5
SMHS CA	93.2	85.7	93.1
SMHS MI	93.6	94.4	93.6
SMHS NY	95.6	90.0	95.6
SMHS TX	94.9	84.6	94.9
Supplemental Security			
HHS	94.7	82.0	94.4
SMHS CA	92.8	81.9	90.6
SMHS MI	93.3	85.4	92.1
SMHS NY	96.1	82.8	94.0
SMHS TX	94.2	88.9	93.0
Social Security			
HHS	94.6	82.1	92.8
SMHS CA	93.1	82.4	90.9
SMHS MI	93.3	81.7	91.1
SMHS NY	95.7	83.4	93.8
SMHS TX	94.2	86.3	92.8

*Logical imputation included in these rates.

Income Type and Sample	Income Amount		
	Zero	Nonzero	Total
Public Assistance			
HHS	94.7	89.9	94.4
SMHS CA	93.6	85.3	91.1
SMHS MI	93.8	84.0	89.7
SMHS NY	94.4	89.7	92.3
SMHS TX	95.1	90.3	93.7
Pension			
HHS	94.5	83.0	93.9
SMHS CA	93.2	88.6	93.1
SMHS MI	93.5	82.6	93.3
SMHS NY	95.6	76.1	95.3
SMHS TX	94.9	84.3	94.8
Cash Payments			
HHS	94.4	78.0	94.1
SMHS CA	93.1	71.0	92.7
SMHS MI	93.5	75.9	93.4
SMHS NY	95.6	73.7	95.3
SMHS TX	94.7	69.0	94.3
Interest			
HHS	88.2	52.0	76.5
SMHS CA	89.4	68.8	88.2
SMHS MI	91.4	65.9	90.0
SMHS NY	93.2	61.3	92.0
SMHS TX	93.6	69.2	93.2
Investment Income			
HHS	93.8	57.5	91.2
SMHS CA	93.1	65.8	92.8
SMHS MI	93.4	77.8	93.3
SMHS NY	95.5	90.9	95.5
SMHS TX	94.9	90.0	94.8
Other Sources			
HHS	94.4	82.3	93.9
SMHS CA	93.1	69.4	92.6
SMHS MI	93.4	84.9	93.0
SMHS NY	95.5	73.1	95.2
SMHS TX	94.7	78.4	94.5

First, missing data were replaced for the variables indicating receipt/lack of receipt of the various forms of income. Having developed imputation-revised income indicators, the next step involved using these variables to replace missing data for those remaining individuals with some income data still missing. When "No" was imputed for an income indicator, the relevant income recode was set to zero. Finally, for individuals with missing nonzero income amounts, the individuals were grouped into classes based upon types of income received and other auxilary data, and then each individual with missing data was linked to an individual with totally complete data using a weighted sequential hot deck algorithm. Then the missing income recode data (and only the missing data) for the individual with partial data were replaced by the recode value found for the donor.

Because many patterns of nonresponse had no donors but only individuals with partially missing data, the imputation was done separately within three income groups: public, private non-interest, and interest income. For public income imputation, classing was by the types of public income received and sorting by sex, race, and employment income. For private income imputation, classing was by the types of private non-interest income received and sorting by sex, race, a collapsed employment income variable, and age. For interest imputation, collapsed versions of age, employment income, sex, race, and education were used for classing with sorting by the uncategorized employment income.

When this imputation step was completed, all individuals had a value defined for each income recode. For the national sample, 56.7

percent provided data for all 12 items and 88.4 percent provided data for at least 11 items. Over the four State Medicaid samples, 74.6 to 81.5 percent provided data for all 12 income items and 91.2 to 94.2 provided data for at least 11 of the 12 items.

Table 9-4 provides imputation results for the individual income items. Considering that approximately five percent of each sample dropped out of the survey before Round 5 when income data were collected, these response rates are quite good. In reporting response for zero versus nonzero income amounts, the imputation-revised data were used to partition individuals as to their income. Thus, 99.8 percent of the national household sample (HHS) with zero as their imputation-revised employment income provided data and 80.7 percent of the individuals with nonzero employment income provided data. One general pattern that emerges from an examination of Table 9-4 is the fact that nonresponse is much lower when individuals have no income to report. The most difficult income types to report are those that vary with time such as employment, interest, and investment income.

9.7 Concluding Remarks

When performing analyses which involve the use of imputation-revised data, the researcher is advised to study the imputation specifications to determine in what ways, if any, the methods used to replace missing data will affect analysis. The methods used to replace missing data should be selected to reduce the nonresponse bias and to minimize the variance induced by imputation. In making inferences based upon imputed data, the effect of nonresponse bias remaining _after_ imputation needs to be considered and the increased variability induced by the imputation. When the response rate is large, both of these effects should have negligible impact. As the response rate decreases, these effects will assume greater importance.

It is especially unfortunate that there is no readily available method of estimating the variance of statistics derived from imputed data. Typically, analysts ignore the fact that imputation was used and compute variance in the usual manner. When the rate of missing data is low, the variance estimates should be affected only negligibly by imputation. For variables where the rate of missing data is high, NMCUES used the weighted sequential hot deck approach which provides more control over the variability induced by imputation.

References

Cox, B. G. (1983). Compensating for Missing Data in the NMCUES. _Priorities in Health Statistics 1983_, DHHS Publication No. (PHS) 81-1214, National Center for Health Statistics, Hyattsville, Maryland, 186-191.

Cox, B. G., A. E. Parker, S. S. Sweetland, and S. C. Wheeless (1982). _Imputation of Missing Item Data for the National Medical Care Utilization and Expenditure Survey_, RTI Report No. RTI/1898/06-02F, Contract No. HRA-233-79-2032, Health Care Financing Administration, Baltimore, Maryland and National Center for Health Statistics, Hyattsville, Maryland.

10
An Analysis of Alternative Attrition Compensation Strategies

Surveys that are characterized by longitudinal designs are subject to two distinct forms of nonresponse. The more traditional form encountered in sample surveys is referred to as complete nonresponse and occurs when an eligible sampling unit, or a member of the unit, does not participate in the survey. In the National Medical Care Expenditure Survey, households and their members were defined as complete nonrespondents when one of the following descriptions was appropriate:

- a dwelling unit determined to be occupied but where no contact could be made with any household members,
- an occupied dwelling unit where contact was made but not with a member at least 14 years of age,
- an occupied dwelling unit where contact was established with a respondent who refused to be interviewed or where an insurmountable language barrier existed, and
- an occupied dwelling unit where the screening information was obtained but the members subsequently refused to complete the questionnaire.

In some instances, responses were obtained from one or more residents of a dwelling unit but not from all residents of the unit. These nonresponding individuals, within otherwise responding households, were also classified as complete nonrespondents. Sample weighting procedures to compensate for complete nonresponse are presented in Chapter 7.

In addition to complete nonresponse, longitudinal surveys also encounter _partial response_ (or _partial nonresponse_), when initially responding survey participants provide data for only part of the time they are eligible to respond. For example, initially responding persons sometimes refuse to participate in later rounds of data collection. Similiarly, inability to locate participants who change residences also results in partial response. The loss of later interviews for initially responding individuals is commonly referred to as _sample attrition_.

For the NMCES, approximately 9 percent of those sampling units eligible for the sample refused to respond; of the remainder, approximately 11 percent of the survey participants did not provide data for the entire period in 1977 that they were eligible to provide data.* Compensation for partial response to the NMCES applies to those data collected over time (e.g., health care utilization and expenditure measures). In this chapter, alternative attrition compensation procedures are described. An investigation is pre-

---

*NMCES sample individuals were eligible to provide data during the time period in 1977 in which they were civilian, noninstitutionalized residents of the United States.

sented of three approaches and their suitability for implementation in the NMCES with an assessment of their effect in reducing nonresponse bias (Cohen, 1982).

10.1 Attrition Compensation Alternatives

To minimize nonresponse bias in panel estimates, an appropriate strategy is needed for those sample members who do not provide data for the entire reference period. Several methods are available to deal with partial response, including direct replacement procedures as well as estimation approaches.

When the rate of partial response is low, it may be preferable to treat partial respondents as complete nonrespondents. Using this approach, only those sample members providing complete data would be included in analyses. When this approach is used, analysis weights must be developed with a nonresponse adjustment to compensate for the fact that complete and partial nonrespondents are not included in analyses. The weight adjustment should be made within weighting classes that are formed to minimize nonresponse bias. That is, classes should be formed that are internally homogeneous and for which the response rate differs between classes. In this chapter, we will refer to this approach as the weight adjustment strategy. The assumption underlying the weight adjustment strategy is that nonrespondents (complete and partial) tend to have survey responses similar to those of complete respondents from the same class.

A second approach is to replace or impute the data that are missing for partial respondents using data provided by complete respondents. With this approach, the partial respondent is first linked to a complete respondent with similar demographic and health care profiles (Ernst, 1978 and Spiers and Knott, 1969). For time periods that the partial respondent failed to respond, the data provided by the complete respondent is copied and imputed to the partial respondent. This approach involves (1) determining the exact time period for which the partial respondent has data missing or the nonresponding time period, (2) linking the partial respondent to a complete respondent, (3) copying the data that the complete respondent provided for the nonresponding time period, and (4) imputing the copied data to the partial respondent. This approach will be referred to as the linked respondent imputation strategy in this chapter. The assumption underlying this imputation strategy is that the data that the partial respondents fail to provide will be similar, on the average, to the data of the complete respondents to whom they are linked.

A third approach is to use the data that the partial respondent did provide to predict their missing data. In effect, this approach adjusts the data that the respondent provided to account for time periods for which the respondent failed to provide data (Kalton, et al, 1981). The adjustment may be directly made by a modification to the analysis weight or indirectly made as a part of estimation. Collectively, we will refer to this approach as the data adjustment strategy. The assumption underlying the data adjustment strategy is that the data that the partial respondent fails to provide can best be predicted by the data that are provided.

To illustrate the data adjustment strategy, let S represent the number of sample respondents (complete and partial), dr(i) represent the number of days for which the i-th respondent (i=1,2,...,S) provided data, and de(i) represent the number of days for which the i-th respondent was eligible to provide data. Using this notation, the data adjustment strategy would estimate the mean expenditures per person as

$$\bar{y} = \sum_{i \varepsilon S} W(i)\, \hat{Y}(i) / \sum_{i \varepsilon S} W(i) \qquad (10\text{-}1)$$

where W(i) is the analysis weight and $\hat{Y}(i)$ is the estimated annual expenditures of the i-th respondent. The latter estimate is obtained as

$$\hat{Y}(i) = [de(i) \,/\, dr(i)]\, Z(i) \qquad (10\text{-}2)$$

where Z(i) is the total expenditures that the i-th respondent reported in the dr(i) days that he responded. When the respondent responds for the entire time period in which he is eligible [that is, dr(i) = de(i)], annual expenditures are known rather than estimated and Y(i) = Z(i). When the respondent provides only partial data, the partial data are weighted up by the reciprocal of the fraction of time responding to obtain an estimate of annual expenditures.

A final approach is to replace or impute for the partial respondent's missing data using data that the partial respondent provided. Thus, if the partial respondent provided data for one-fourth of the year, three copies of the data would be created and used to replace the missing data. We will refer to this approach as the linked-self imputation strategy. The underlying assumption behind the strategy is the same as the assumption underlying the data adjustment strategy - that data missing for partial respondents are best predicted by the data provided during responding time periods.

10.2 Characteristics of NMCES Complete and Partial Respondents

Of the 38,815 participants in the NMCES survey, 10.7 percent failed to respond for the entire time for which they were eligible in 1977. To develop strategies to compensate for the lost data of partial respondents, characteristics of the population were examined to identify systematic differences from the population of complete respondents. To make these comparisons, demographic profiles of these two distinct populations were computed for self-reported health status, race, sex, age, years of school completed, marital status, and size of city. The estimated population distributions for the two respondent groups can be observed in Table 10-1. Studentised differences were computed to determine whether significant differences existed in the demographic configurations of the two respondent groups.

The comparisons of marital status distributions for the two respondent groups revealed a significantly larger percentage of complete respondents who were married, and a correspondingly larger percentage of divorced or never married individuals among the partial respondents. There was also a significantly larger percentage of the partial respondents living in the 16 largest Standard Metropolitan Statistical Areas (SMSAs). Complete respondents were more likely to assess their health as excellent as opposed to the partial respondents, for whom a larger percentage were in poor health. Finally, a larger proportion of the partial respondents were nonwhite than the complete respondents.

Cross-classifications by race (Table 10-2) indicated that complete respondents were more likely to be whites in excellent health than the partial respondents, while a larger percentage of partial respondents were nonwhites in good or excellent health. Complete respondents had a significantly higher relative frequency of whites between the ages of 25 to 44 than partial nonrespondents, whereas the partial respondents demonstrated a larger percentage of nonwhites less than five years old. In addition, complete respondents were more likely to be married whites than the partial respondents, who had a greater percentage of divorced whites and married nonwhites.

Table 10-1. Comparison of Demographic Profiles for Complete and Partial Respondents to the NMCES

Demographic Variable	Complete Respondents		Partial Respondents		Studentised Difference
	Proportion	Standard Error	Proportion	Standard Error	
Health Status:					
Excellent	0.449	0.007	0.392	0.013	3.914*
Good	0.371	0.005	0.381	0.012	-0.794
Fair	0.103	0.003	0.108	0.007	-0.571
Poor	0.033	0.002	0.042	0.004	-1.961*
Unknown	0.045	0.002	0.077	0.008	-3.881*
Race:					
White	0.872	0.010	0.828	0.014	2.509*
Nonwhite	0.128	0.010	0.172	0.014	-2.509*
Sex:					
Male	0.482	0.002	0.484	0.006	-0.170
Female	0.518	0.002	0.516	0.006	0.170
Years of Education:					
0 to 8 years	0.109	0.004	0.107	0.006	0.264
9 to 11 years	0.128	0.003	0.133	0.007	-0.615
12 years	0.242	0.003	0.226	0.009	1.573
13 to 15 years	0.107	0.003	0.102	0.006	0.642
16 or more years	0.090	0.004	0.080	0.006	1.398
17 years of age or unknown education	0.324	0.005	0.352	0.009	-2.720*

Marital Status:					
Under 17 years of age	0.282	0.003	0.284	0.009	-0.244
Never married	0.131	0.003	0.148	0.007	-2.148*
Married	0.447	0.004	0.408	0.009	4.008*
Widowed	0.053	0.002	0.060	0.004	-1.402
Separated	0.019	0.001	0.025	0.003	-1.899
Divorced	0.033	0.001	0.042	0.004	-2.109*
Unknown	0.035	0.001	0.033	0.004	0.485
Size of City:					
SMSA: 16 largest SMSAs	0.247	0.020	0.351	0.312	-2.751*
SMSA: 500,000 or more population but not 16 largest	0.254	0.029	0.268	0.033	-0.310
SMSA: less than 500,000 population	0.186	0.031	0.143	0.028	1.011
Not SMSA: less than 60% rural	0.190	0.022	0.145	0.021	1.466
Not SMSA: 60% or more rural	0.123	0.021	0.093	0.030	0.835
Age:					
0 to 4 years	0.072	0.002	0.074	0.005	-0.467
5 to 14 years	0.171	0.003	0.174	0.008	-0.317
15 to 24 years	0.186	0.003	0.198	0.008	-1.418
25 to 34 years	0.153	0.003	0.144	0.007	1.132
35 to 44 years	0.111	0.002	0.096	0.005	2.603*
45 to 54 years	0.108	0.002	0.117	0.005	-1.607
55 to 64 years	0.095	0.002	0.094	0.006	0.111
65 years and up	0.105	0.004	0.103	0.007	0.253

*Denotes statistically significant difference at the $\alpha = 0.05$ level.

Table 10-2. Comparison of Demographic Profiles for Complete and Partial NMCES Respondents by Race

Demographic Variable	Complete Respondents		Partial Respondents		
	Proportion	Standard Error	Proportion	Standard Error	Studentised Difference
Race and Health Status:					
White:					
Excellent	0.408	0.008	0.335	0.014	4.470*
Good	0.315	0.005	0.304	0.012	0.821
Fair	0.084	0.003	0.087	0.006	-0.328
Poor	0.026	0.001	0.036	0.004	-2.181*
Unknown	0.038	0.002	0.066	0.009	-3.051*
Nonwhite:					
Excellent	0.041	0.004	0.057	0.007	-2.023*
Good	0.056	0.005	0.077	0.007	-2.466*
Fair	0.019	0.002	0.021	0.003	-0.598
Poor	0.006	0.001	0.005	0.001	0.560
Unknown	0.007	0.001	0.011	0.003	-1.712
Race and Age:					
White:					
0 to 4 years	0.060	0.002	0.054	0.005	1.130
5 to 14 years	0.143	0.003	0.141	0.009	0.176
15 to 24 years	0.160	0.003	0.162	0.008	-0.248
25 to 34 years	0.135	0.003	0.113	0.007	2.874*
35 to 44 years	0.097	0.002	0.077	0.006	3.403*
45 to 54 years	0.096	0.002	0.104	0.005	-1.461

55 to 65 years	0.086	0.002	0.082	0.006	0.634
65 years and older	0.095	0.003	0.094	0.007	0.093
Nonwhite:					
0 to 4 years	0.012	0.001	0.020	0.003	-2.514*
5 to 14 years	0.028	0.002	0.032	0.004	-0.936
15 to 24 years	0.026	0.002	0.036	0.004	-2.057*
25 to 34 years	0.018	0.002	0.031	0.004	-2.782*
35 to 44 years	0.013	0.001	0.019	0.002	-2.391*
45 to 54 years	0.012	0.001	0.013	0.002	-0.199
55 to 65 years	0.009	0.001	0.012	0.002	-1.466
65 years and older	0.010	0.001	0.009	0.002	0.547
Race and Marital Status:					
White:					
Under 17 years of age	0.236	0.005	0.226	0.011	0.937
Never married	0.111	0.003	0.122	0.007	-1.431
Married	0.407	0.006	0.349	0.011	4.868*
Widowed	0.046	0.002	0.051	0.004	-1.234
Separated	0.042	0.002	0.044	0.004	-0.354
Divorced	0.028	0.001	0.037	0.004	-2.193*
Nonwhite:					
Under 17 years of age	0.046	0.004	0.059	0.007	-1.721
Never married	0.020	0.002	0.026	0.003	-1.861
Married	0.040	0.003	0.059	0.006	-2.948*
Widowed	0.007	0.001	0.009	0.002	-0.856
Separated	0.011	0.001	0.014	0.002	-1.143
Divorced	0.005	0.001	0.005	0.001	-0.086

*Denotes statistically significant difference at the $\alpha = 0.05$ level.

10.3 Experimental Approach to Attrition Compensation

In the NMCES, a controlled experiment was conducted to determine the best method of compensation for the lost data of partial respondents. The three alternative strategies under consideration included: (1) the weight adjustment strategy that uses data from complete respondents only, (2) the imputation strategy that substitutes data from a linked complete respondent who matched the partial respondent on demographic characteristics, and (3) the data adjustment strategy that weights up respondent data to account for missing time periods. Cost considerations associated with implementing more than one imputation procedure, together with the recognition that the linked-self imputation strategy results would be similar to the data adjustment approach, led to the decision to exclude the linked-self imputation approach from the investigation.

To evaluate the three selected strategies, partial data were synthetically created for individuals with complete information, to create a test data set exhibiting partial nonresponse similar to that the NMCES, but for which the missing responses would be known. The availability of complete data for the synthetically-derived partial respondents allowed calculation of the effect on survey estimates of the alternative attrition compensation strategies.

The study was based upon expenditure and utilization data for prescribed medicines from the NMCES. Prescribed medicines were selected for the study due to their wide user population, unlike hospital visits which are made by a much smaller segment of the national population. In addition, the relationship between prescribed medicine usage and health insurance coverage and the positive correlation with medical provider visit data argued for the generalizability of study findings (Aday and Eichhorn, 1972; Phelps and Newhouse, 1974).

The process of synthetically transforming a subset of the complete respondents into partial respondents was initiated by matching the 4,146 partial respondents to the 34,669 complete respondents on relevant demographic measures. The comparison of the partial and complete respondents revealed significant differences in population composition for the two groups. To preserve the characteristics of the true partial respondents when forming the synthetic partial respondents from complete respondents, matching was done on the following characteristics: self-reported health status, age, race, sex, Census region (Northeast, North Central, South, West), insurance coverage (Medicaid coverage, private insurance and not Medicaid, all others) and the number of prescribed medicines obtained for the time period the partial respondent provided data. The last two matching variables were used to insure consistency in expenditure, utilization, and source of payment data between true and synthetic partial respondents.

The synthetic derivation of partial respondent data was implemented in the following manner. True partial respondents were linked to complete respondents based upon the matching variables specified above. Data for the matched complete respondents were then removed for the time period the partial respondent had not provided data. The result is synthetic partial data for a person whose true complete data is known.

The first method investigated was the weight adjustment approach which used only the data from the synthetically-designated complete respondents to characterize the nation. In this setting, the 4,146 synthetic partial individuals were viewed as total nonrespondents and a standard nonresponse adjustment was made that implicitly attributed the characteristics of respondents within the same age-race-sex weighting class to nonrespondents (Chapman, 1976). Age, race, and sex were used in forming weighting classes since utilization and expenditure measures are strongly associated with these variables and age, race, and sex were used in the NMCES weight adjustment for total nonresponse.

Within weighting classes, the weights of the 30,523 designated complete respondents were inflated by the nonresponse adjustment factor,

$$A(c) = \sum_{i \varepsilon T(c)} W(i) \Big/ \sum_{i \varepsilon R(c)} W(i) \qquad (10\text{-}3)$$

where W(i) is the total-nonresponse-adjusted analysis weight for the i-th respondent, the summation in the numerator is over the total group of respondents in the c-th weighting class, and the summation in the denominator is over complete respondents in the c-th weighting class. The partial-nonresponse-adjusted weight for complete respondents took the form

$$\omega(i) = A(c)\ W(i) \qquad (10\text{-}4)$$

These adjusted weights were then used with the data of the complete respondents in calculating parameter estimates.

The second strategy investigated was the linked-respondent imputation strategy. For simplicity's sake, we will refer to the second approach as the imputation strategy in the remainder of this chapter. To implement the imputation strategy, the 4,146 synthetically-designated partial respondents were matched to the remaining 30,523 complete respondents based upon the same set of characteristics used in the earlier matching (i.e., self-reported health status, age, race, sex, Census region, insurance coverage, and number of prescribed medicines obtained for the time period the partial respondent provided data). Once a synthetic partial respondent was linked to a complete respondent, prescribed medicine events made by the complete respondent were imputed to the partial respondent when they occurred during the nonresponding time period of the partial respondent. Parameter estimates were then derived using the imputation-revised data in conjunction with the analysis weights associated with complete and partial respondents. This strategy is similar to the hot deck technique used to adjust for item nonresponse, which was discussed in Chapter 8.

The third strategy investigated was a variation of the data adjustment strategy. In the modified strategy, data adjustments were computed separately within adjustment classes defined by the cross-classification of age, race, and sex and an indicator of complete versus partial response. Adjustment within classes was preferred over a person-specific data adjustment, to avoid excessively large inflation factors for partial data. The implicit assumption being made is that the utilization experience of partial respondents within each adjustment class is uniform across their periods of response and nonresponse.

For each respondent, the number of days the respondent was eligible for the study was determined [de(i) for the i-th respondent], and the number of eligible days for which the participant responded [dr(i) for the i-th respondent]. Within each of the adjustment classes (c), the respondents' analysis weights [W(i)] were used in conjunction with the de(i) and dr(i) factors to compute a data adjustment factor, D(c), which took the form:

$$D(c) = \sum_{i \varepsilon c} W(i)\, de(i) \Big/ \sum_{i \varepsilon c} W(i)\, dr(i) \qquad (10\text{-}5)$$

where the summation is over all respondents within the c-th adjustment class. For the complete respondents, the adjustment factor was equal to one [D(c) = 1]. Each respondent within adjustment class c then had the time dependent data that he had provided, Z(i), inflated by the factor D(c) to yield an adjusted value:

$$\hat{Y}(i) = Z(i)\, D(c) \qquad (10\text{-}6)$$

Parameter estimates were then computed using the respondents' analysis weights and these adjusted data values.

10.4 Analytical Results

Event and person level analyses of prescribed medicines usage and expenditures were used to compare the alternative strategies. Specifically, the following criterion variables were estimated from the prescribed medicine data:

- mean charge per prescribed medicine event,
- mean charge per event by source of payment (family, private insurance, Medicaid, Medicare, other),
- event-level distribution of total charge for prescribed medicines, categorized by classes: \$0.00 to \$4.99, \$5.00 to \$9.99, \$10.00 to \$14.99, \$15.00 to \$19.99, and \$20.00 and up,
- mean expense per person with an event,
- mean expense per person with an event by source of payment,
- mean number of prescribed medicines per person with an event, and
- person-level distribution by number of prescribed medicines, categorized by classes: 0, 1, 2 to 3, 4 to 6, and 7 and up.

Estimates were computed for domains defined by the following variables: self-reported health status, Census region, age, race, and sex. Forty-one distinct domains were created by the cross-classification of age, race, and sex (32 domains) and the univariate distributions of health status (4 domains) and region (4 domains), in addition to the total population domain.

Parameter estimates computed from the full set of data for the complete respondents served as a truth set to which estimates generated through the alternative strategies could be compared. Hereafter, these estimates will be referred to as the "complete data estimates." To test for statistical equivalence between the complete data estimates and the estimates resulting from the alternative compensation strategies, Studentised statistics were computed for the 41 domains specified above. In this setting, the complete data estimate was treated as a population parameter and a 0.05 level of significance was used for hypothesis testing. The test statistic took the form

$$t = \frac{(\bar{y}_{acs} - \bar{y}_{comp})}{[Var(\bar{y}_{acs})]^{\frac{1}{2}}} \tag{10-7}$$

where $\bar{y}_{acs}$ is the mean estimate derived by one of the attrition compensation strategies and $\bar{y}_{comp}$ is the complete data estimate. For the sake of convenience, the variance of $\bar{y}_{acs}$ was estimated in the usual manner by a Taylor Series linearization method (Woodruff, 1971 and Shah, 1981) rather than by replication of the attrition compensation process. This variance estimate includes variance due to the NMCES area sampling as well as the attrition compensation process. Consequently, it is a conservative estimate for hypothesis testing.*

For event-level analyses, none of the differences between the complete data estimate and each of the three strategy estimates was significantly different from zero for any of the 41 domains. The statistics examined were mean charge per prescribed medicine, mean charge per prescribed medicine by source of payment, and the distribution of charge for prescribed medicines. Estimates of these prescribed medicine event-level estimates for the total population, derived from the attrition compensated data and from the complete

---

*Since the appropriate variance estimate will be much smaller (i.e., will contain only the component due to attrition compensation), fewer hypotheses will be rejected at the five percent level of significance than should be when testing at the $\alpha = 0.05$ level of significance.

data, are displayed in Table 10-3 to illustrate the degree of uniformity across methodologies.

No significant differences were detected when the 41 complete data domain estimates were compared with the attrition compensation strategy estimates of the mean expense per person with prescribed medicines, and the mean expense by source of payment. Similarly, the comparative tests relative to person-level utilization estimates, expressed in terms of domain means, revealed no significant differences in parameter estimates of prescribed medicine use. The total population estimates for these person-level prescribed medicine measures are displayed in Table 10-4 to again indicate points of the similarity between the three strategy estimates and the complete data estimate.

For the weight adjustment strategy and the imputation strategy, no significant differences from the complete data estimates were observed across the 41 domains for the person-level distribution by annual number of prescriptions. Significant differences did occur between the data adjustment strategy estimate and the complete data estimate for these person-level distributional comparisons of prescribed medicine use. Domain estimates of the proportion of the population with no prescribed medicines were significantly higher for the data adjustment strategy. This was determined by observing 18 significant Studentised differences among the 41 domain comparisons, which is much more than the 2 significant values that would be expected to occur by chance. The data adjustment strategy was also observed to yield lower estimates of individuals with only one pre-

Table 10-3. Comparison of Total Population Event-Level Estimates for the Alternative Compensation Strategies to the Complete Data Estimates

Health Care Parameter	Estimated Value and Standard Error			
	Weight Adjustment Strategy	Imputation Strategy	Data Adjustment Strategy	Complete Data Estimate
Mean Charge Per Prescribed Medicine (in dollars)	6.062 (0.050)	6.074 (0.046)	6.046 (0.049)	6.075 (0.048)
Mean Charge Per Prescribed Medicine by Source of Payment (in dollars):				
Family	4.367 (0.058)	4.395 (0.055)	4.378 (0.057)	4.398 (0.058)
Private Insurance	0.839 (0.045)	0.845 (0.040)	0.834 (0.045)	0.832 (0.043)
Medicaid	0.496 (0.034)	0.474 (0.033)	0.478 (0.034)	0.482 (0.034)

Medicare	0.017 (0.004)	0.015 (0.003)	0.016 (0.004)	0.017 (0.004)
All Other	0.343 (0.033)	0.344 (0.032)	0.340 (0.033)	0.347 (0.033)
Proportional Distribution of Charge for Prescribed Medicines				
$ 0.00 to $ 4.99	0.494 (0.005)	0.493 (0.005)	0.496 (0.005)	0.493 (0.005)
$ 5.00 to $ 9.99	0.387 (0.005)	0.387 (0.004)	0.385 (0.004)	0.387 (0.004)
$10.00 to $14.99	0.084 (0.002)	0.086 (0.002)	0.086 (0.002)	0.086 (0.002)
$15.00 to $19.99	0.021 (0.001)	0.021 (0.001)	0.021 (0.001)	0.021 (0.001)
$20.00 and up	0.013 (0.001)	0.013 (0.001)	0.013 (0.001)	0.013 (0.001)

Table 10-4. Comparison of Total Population Person-Level Estimates for the Alternative Compensation Strategies to the Complete Data Estimates

Health Care Parameter	Estimated Value and Standard Error			
	Weight Adjustment Strategy	Imputation Strategy	Data Adjustment Strategy	Complete Data Estimate
Mean Expense Per Person - Persons with Prescribed Medicine (in dollars)	43.309 (0.889)	43.344 (0.904)	44.951 (0.910)	43.546 (0.889)
Mean Expense Per Person by Source of Payment - Persons with Prescribed Medicines in dollars):				
Family	31.204 (0.643)	31.362 (0.647)	32.548 (0.653)	31.524 (0.647)
Private Insurance	5.995 (0.352)	6.030 (0.326)	6.203 (0.369)	5.963 (0.349)
Medicaid	3.543 (0.268)	3.385 (0.258)	3.554 (0.276)	3.456 (0.266)

Medicare	0.119 (0.027)	0.110 (0.024)	0.122 (0.027)	0.120 (0.026)
All Other	2.448 (0.237)	2.457 (0.231)	2.525 (0.251)	2.484 (0.240)
Mean Number of Prescribed Medicines - Persons with Prescribed Medicines	7.145 (0.125)	7.136 (0.127)	7.435 (0.129)	7.168 (0.117)
Person-Level Proportional Distribution by Number of Prescribed Medicines:				
0	0.388 (0.005)	0.391 (0.005)	0.418 (0.005)	0.392 (0.005)
1	0.134 (0.002)	0.133 (0.002)	0.118 (0.002)	0.134 (0.002)
2 to 3	0.165 (0.003)	0.166 (0.003)	0.160 (0.003)	0.164 (0.003)
4 to 6	0.122 (0.002)	0.121 (0.002)	0.117 (0.002)	0.120 (0.002)
7 or more	0.190 (0.004)	0.189 (0.004)	0.188 (0.004)	0.190 (0.004)

scribed medicine. This was determined by noting 18 significant differences for the 41 comparisons of domain estimates. For the remaining categories of the utilization distribution, specifically the classes: 2 to 3, 4 to 6, and 7 or more prescriptions, estimates derived by the data adjustment strategy were found to be statistically equivalent to the distribution derived from complete respondents.

The higher estimated population proportion with no prescribed medicines and the lower estimated proportion with one prescription that were observed for the data adjustment strategy should be expected. By the nature of the adjustment procedure, partial respondents with no prescribed medicines reported for the period they responded would automatically be classified as having zero medicines (i.e., the product of zero medicines and an adjustment factor greater than one remains zero). However, persons that have only one prescribed medicine during the year will have more time periods in which no medicines were obtained than time periods with one medicine. When the time period(s) of no medicine obtained corresponds with the time for which the partial respondent provides data, the partial respondent will be classified as having zero medicines under the data adjustment strategy.

10.5 Estimation of Bias and Mean Square Error for the Alternative Strategies

In considering the implementation of a nonresponse compensation procedure, the quality of the resultant statistics is of primary concern. Since nonresponse compensations are made to reduce the bias associated with systematic differences between respondents and nonrespondents, the amount of bias remaining in survey estimates after use of a compensation procedure is crucial. Normally, the bias cannot be determined in any useful analytic fashion since it will depend upon whether the assumptions hold that underlie the compensation method. For instance, all of the strategies described in this chapter assume that complete and partial respondents within classes have responses that are similarly distributed. The extent to which complete and partial respondents differ within classes will determine the amount of bias in the resulting estimators. Other factors of importance are the degree to which responses are homogeneous within classes and the extent to which complete data rates differ between classes.

By using synthetically derived partial data, the bias associated with the three strategy estimates could easily be estimated as the difference between the strategy estimate $[\hat{\mu}_{s(i)}]$ and the complete data estimate $[\hat{\mu}_c]$, or

$$\text{Bias}[\hat{\mu}_{s(i)}] \doteq \hat{\mu}_{s(i)} - \hat{\mu}_c \tag{10-8}$$

In this context, the mean square error (MSE) of the strategy estimate $[\hat{\mu}_{s(i)}]$ is defined to be the expected value of the squared difference between the strategy i estimate and the complete data estimate and may be estimated as

$$\text{MSE}[\hat{\mu}_{s(i)}] \doteq [\hat{\mu}_{s(i)} - \hat{\mu}_c]^2 + \text{Var}[\hat{\mu}_{s(i)}] \tag{10-9}$$

The ultimate problem for the data analyst is to construct sample estimates in such a way that the total error of the results is minimized (not just the random sampling error). The mean square error is frequently used as a measure of total error.

Before preceding, it should be noted that the latter estimate of the mean square error of the strategy estimate will be biased to the extent that the correlation between $\hat{\mu}_c$ and $\hat{\mu}_{s(i)}$ is different from unity. The reason for this is that

$$E[\hat{\mu}_{s(i)} - \hat{\mu}_c]^2 = \{\text{Bias}[\hat{\mu}_{s(i)}]\}^2 + \text{Var}[\hat{\mu}_{s(i)}] + \text{Var}[\hat{\mu}_c] - 2\ \text{Cov}[\hat{\mu}_{s(i)}, \hat{\mu}_c] \tag{10-10}$$

Since $\hat{\mu}_{s(i)}$ and $\hat{\mu}_c$ are estimated from largely the same data set (except for moderate nonresponse), one would expect their correlation to be close to unity. With this anticipated high correlation, the joint contribution of the extra terms

$$\text{Var}[\hat{\mu}_{s(i)}] + \text{Var}[\hat{\mu}_c] - 2\ \text{Cov}[\hat{\mu}_{s(i)}, \hat{\mu}_c]$$

should be small.

In most surveys, the bias of sample estimates is not known. However, in this study, the bias remaining in survey estimates after use of a compensation strategy could be determined since the complete data were also available. Considering the fact that 13 percent of the complete respondents were synthetically designated as partial respondents and had data removed accordingly, the overall quality of all three strategy estimates appears to be relative good. Three measures of data quality - the relative bias, the relative root mean square error, and the bias ratio - were used to evaluate the accuracy of total population estimates derived using the three attrition compensation strategies.

The relative bias is defined to be the bias divided by the population value being estimated. This measure of data quality reflects the common specification of data analysts that greater accuracy is needed for estimates of small population values. For this study, the relative bias of a strategy estimate was estimated as the bias (the difference between the statistic and the complete data estimate) divided by the complete data estimate expressed as a percentage (Table 10-5). Of the 23 parameters estimated in this study, all of the relative biases associated with the weight adjustment strategy are moderate (between -5 percent and +5 percent). Two parameter estimates derived from the imputation strategy have larger relative biases, the statistics related to Medicare utilization and expenditures. Three of the data adjustment parameter estimates have larger relative biases, the Medicare utilization estimate and the person-level estimates of the proportion with zero and one medicines prescribed during 1977.

A better measure of the quality of the strategy estimates is furnished by the relative root mean square error, which was defined in this study to be the square root of the mean square error of the strategy estimate divided by the complete data estimate, expressed as a percentage. The relative root mean square error for the three strategy estimates and the complete data estimate are displayed in Table 10-6. (Under the definition used in this study, the bias of the complete data estimate is zero, so the standard error and root

Table 10-5. Comparison of Relative Bias for the Total Population Estimates Obtained Using the Alternative Compensation Strategies

Health Care Parameter	Complete Data Estimate	Relative Bias (%)		
		Weight Adjustment Strategy	Imputation Strategy	Data Adjustment Strategy
Mean Charge Per Prescribed Medicine (in dollars)	6.075	-0.21	-0.02	-0.48
Mean Charge Per Prescribed Medicine by Source of Payment (in dollars):				
Family	4.398	-0.70	-0.07	-0.45
Private Insurance	0.832	0.84	1.56	0.24
Medicaid	0.482	2.90	-1.66	-0.83
Medicare	0.017	0.00	-11.76	-5.88
All Other	0.347	-1.15	-0.86	-2.02
Proportional Distribution of Charge Relative to Events				
$ 0.00 to $ 4.99	0.493	0.20	0.00	0.61
$ 5.00 to $ 9.99	0.387	0.00	0.00	-0.52
$10.00 to $14.99	0.086	-2.33	0.00	0.00
$15.00 to $19.99	0.021	0.00	0.00	0.00

Mean Expense per Person - Persons with Prescribed Medicines (in dollars)	43.546	-0.54	-0.46	3.23
Mean Expense per Person by Source of Payment - Persons with Prescribed Medicines (in dollars)				
Family	31.524	-1.02	-0.51	3.25
Private Insurance	5.963	0.54	1.12	4.02
Medicaid	3.456	2.52	-2.05	2.84
Medicare	0.120	-0.83	-8.33	1.67
All Other	2.484	-1.45	-1.09	1.65
Mean Number of Prescribed Medicines - Persons with Prescribed Medicines	7.168	-0.32	-0.45	3.72
Person-Level Proportional Distribution on Number of Prescribed Medicines				
0	0.392	-1.02	-0.26	6.63
1	0.134	0.00	-0.75	-11.94
2 to 3	0.164	0.61	1.22	-2.44
4 to 6	0.120	1.67	0.83	-2.50
7 or more	0.190	0.00	-0.53	-1.05

Table 10-6. Comparison of Relative Root Mean Square Error for the Estimates Derived from the Complete Data and the Alternative Compensation Strategies

Health Care Parameter	Complete Data	Relative Root Mean Square Error (%)		
		Weight Adjustment Strategy	Imputation Strategy	Data Adjustment Strategy
Mean Charge Per Prescribed Medicine (in dollars)	0.79	0.85	0.76	0.94
Mean Charge Per Prescribed Medicine by Source of Payment (in dollars):				
Family	1.32	1.50	1.25	1.37
Private Insurance	5.17	5.47	5.06	5.41
Medicaid	7.05	7.63	7.04	7.10
Medicare	23.53	23.53	21.21	24.25
All Other	9.51	9.58	9.26	9.72
Proportional Distribution of Charge Relative to Events				
$ 0.00 to $ 4.99	1.01	1.03	1.01	1.18
$ 5.00 to $ 9.99	1.03	1.29	1.03	1.16
$10.00 to $14.99	2.33	3.29	2.33	2.33
$15.00 to $19.99	4.76	4.76	4.76	4.76
$20.00 and up	7.69	7.69	7.69	7.69

Mean Expense per Person - Persons with Prescribed Medicines (in dollars)	2.04	2.11	2.13	3.84
Mean Expense per Person by Source of Payment - Persons with Prescribed Medicines (in dollars)				
Family	2.05	2.28	2.12	3.85
Private Insurance	5.85	5.93	5.58	7.38
Medicaid	7.70	8.15	7.74	8.47
Medicare	21.67	22.52	21.67	22.56
All Other	9.66	9.65	9.36	10.24
Mean Number of Prescribed Medicines - Persons with Prescribed Medicines	1.63	1.77	1.83	4.14
Person-Level Proportional Distribution on Number of Prescribed Medicines				
0	1.28	1.63	1.30	6.75
1	1.49	1.49	1.67	12.03
2 to 3	1.83	1.93	2.20	3.05
4 to 6	1.67	2.36	1.86	3.00
7 or more	2.11	2.11	2.17	2.35

mean square error are identical.) For the complete data estimates, 14 have moderate relative mean square errors (less than five percent). The two largest are associated with estimation of Medicare utilization and expenditure measures. In general, the relative root mean square errors of the three strategy estimates are comparable to those of the complete data estimate. Two exceptions occur for the data adjustment strategy which has a much larger relative root mean square error for the person-level estimates of the proportion of the population with zero and one prescribed medicines in 1977.

Note that these complete data estimates do not reflect the quality of the NMCES data base since they contain only the 89 percent complete respondents. For the full NMCES data base, smaller relative root mean square errors would be anticipated for each of the three estimation strategies. The larger sample size is expected to yield a smaller variance term, which is a component of the numerator of the relative root mean square error. Similarly, the smaller proportion of partial respondents to a larger sample (4,146 out of 38,815), when compared with the experiment sample (4,146 out of 34,669) should also result in a reduction in bias.

A final measure of the quality of the strategy estimates is the bias ratio. Defined as the bias of an estimate divided by the standard deviation of the estimate, the _bias ratio_ provides a measure of the effect of bias on survey inferences. When bias is present, the entire sampling distribution of the statistic $\hat{\theta}$ about the parameter θ is displaced by an amount equal to the bias (Kish, 1965, pp. 566-571). This results in a distortion of the areas in the tails of the sampling distribution of the appropriate test statistic.

For example, suppose $[\hat{\theta}-\text{Bias}(\hat{\theta})-\theta]/S(\hat{\theta})$ is distributed as Student's t where $\text{Bias}(\hat{\theta})$ and $S(\hat{\theta})$ are the bias and standard deviation of $\hat{\theta}$, respectively. In this case, the proper two-tailed confidence limits for θ with confidence coefficient α are given by

$$\hat{\theta} + t_{\alpha} S(\hat{\theta}) [\pm 1 - \text{Bias}(\hat{\theta})/t S(\hat{\theta})]$$

where t_{α} is the value of the t statistic at $1-\alpha/2$. If the bias is positive, the standard interval represented by the confidence limits

$$\hat{\theta} \pm t_{\alpha} S(\hat{\theta})$$

will exceed θ more frequently than it should. Negative bias will have the opposite effect. For each increase of 0.5 in the absolute value of the bias ratio or $|\text{Bias}(\hat{\theta})/S(\hat{\theta})|$, the probability that the corresponding interval fails to contain θ will be approximately double. If, for example, the bias ratio for a sample statistic is 1.0 or -1.0, a presumed 95 percent confidence interval for $\hat{\theta}$ would have a chance of about 20 percent of excluding θ instead of a 5 percent chance.

Since a large sample size results in small variances (34,669 for the data adjustment and imputation strategies and 30,523 for the weight adjustment strategy), even a moderate bias can have a large effect (Table 10-7). Two of the 23 estimates derived from the weight adjustment strategy have absolute bias ratios equal to 1.0 - the statistics for the proportion of prescribed medicines with

Table 10-7. Comparison of Bias Ratios for Estimates Obtained Using the Alternative Compensation Strategies

Health Care Parameter	Bias Ratio		
	Weight Adjustment Strategy	Imputation Strategy	Data Adjustment Strategy
Mean Charge Per Prescribed Medicine (in dollars)	-0.26	-0.02	-0.59
Mean Charge Per Prescribed Medicine by Source of Payment (in dollars):			
Family	-0.53	-0.05	-0.35
Private Insurance	0.16	0.33	0.04
Medicaid	0.41	-0.24	-0.12
Medicare	0.00	-0.67	-0.25
All Other	-0.12	-0.09	-0.21
Proportional Distribution of Charge Relative to Events			
$ 0.00 to $ 4.99	0.20	0.00	0.60
$ 5.00 to $ 9.99	0.00	0.00	-0.50
$10.00 to $14.99	-1.00	0.00	0.00
$15.00 to $19.99	0.00	0.00	0.00
$20.00 and up	0.00	0.00	0.00

Mean Expense per Person - Persons with Prescribed Medicines (in dollars)	-0.27	-0.22	1.54
Mean Expense per Person by Source of Payment - Persons with Prescribed Medicines (in dollars)			
Family	-0.50	-0.25	1.57
Private Insurance	0.09	0.21	0.65
Medicaid	0.32	-0.28	0.36
Medicare	-0.04	-0.42	0.07
All Other	-0.15	-0.12	0.16
Mean Number of Prescribed Medicines - Persons with Prescribed Medicines	-0.18	-0.25	2.07
Person-Level Proportional Distribution on Number of Prescribed Medicines			
0	-0.80	-0.20	5.20
1	0.00	-0.50	-8.00
2 to 3	0.33	0.67	-1.33
4 to 6	1.00	0.50	-1.50
7 or more	0.00	-0.25	-0.50

charges between $10.00 and $14.99, and the proportion of persons with 4 to 6 prescriptions in 1977. The imputation strategy has no estimates with an absolute bias ratio greater than 1.0. The performance of the data adjustment approach is the worst with respect to the bias ratio, with 7 of the 23 estimators having an absolute bias ratio greater than 1.0. The statistics for which the data adjustment strategy have large bias ratios are the mean expense per person with at least one prescribed medicine, the mean expense paid by the family per person with prescribed medicines, the mean number of prescribed medicines per person with prescribed medicines, and the proportions of persons with 0, 1, 2 to 3, and 4 to 6 medicines prescribed in 1977.

The bias ratios presented in Table 10-7, however, are not representative of the bias ratios that would have resulted had the three strategies been applied to the full NMCES data base. In this case, it is not clear which direction the bias ratios will take. It will be a function of the relative magnitude of the expected reduction of both the numerator (the bias) and the denominator (the standard deviation), when compared with the experimental sample results.

The results of these analyses of the relative bias, relative root mean square error, and bias ratio suggest that the imputation approach is probably best in terms of compensating for partial response with the weight adjustment strategy a close second. The deficiencies of the data adjustment approach are most notable for the person-level distributional estimates and suggest that one of the alternate approaches should be used for this type of estimation.

10.6 Concluding Remarks

This controlled study examining the accuracy of three alternative strategies for partial nonresponse compensation was made to determine the method for NMCES use. As observed, the domain estimates of the mean expenditures and utilization derived by the three strategies were reasonably comparable to the complete data estimates. The most noticeable deviation in all comparisons with actual data occurred for person-level distributional estimates of prescribed medicine utilization when the data adjustment strategy was used to compensate for partial data. The data adjustment strategy resulted in overestimating the proportion of the domain populations with no utilization experience while simultaneously underestimating the proportion of the populations with one prescribed medicine event.

For the weight adjustment strategy, which considered the partial respondents as total nonrespondents and inflated the data of complete respondents by a standard nonresponse adjustment, the resultant estimates were reasonably unbiased for most of the prescribed medicine utilization and expenditure measures. However, the loss of an additional 11 percent of the sample due to partial nonresponse was thought to be undesirable for a variety of reasons. The deletion of partial respondents from the NMCES data base requires the implicit assumption that within weighting classes partial respondents and total nonrespondents are both similar to each other and the complete respondents with respect to demographic composition and other utilization, expenditure, and health measures of relevance to the NMCES study. However, the analysis comparing the demographic characteristics of the partial and complete respondents revealed systematic differentials in composition, and it is unlikely that all of these differences could be controlled in defining weighting classes. Although no substantial biasing of the utilization and expenditure estimates for prescribed medicines was observed, other health care estimates, such as physician and hospital utilization and expenditure estimates, might prove more sensitive to the population differentials that distinguish the two respondent subgroups. In addition, the deletion of the partial respondents from the data base limits the capacity for subsequent analyses or methodological investigations that require the inclusion of this subgroup.

Clearly, the imputation strategy to correct for partial nonresponse was best in terms of accuracy and satisfying underlying assumptions of homogeneity inherent in the matching of the partial respondents with their complete counterparts. When time and cost are considered, as was required for the NMCES where cost efficiency was a primary concern, the advantage of this method of imputation for the NMCES was less obvious. In addition to linking partial respondents to complete respondents with similar characteristics, data for the nonresponding time period of the partial respondent would have to be copied from the linked complete respondent and transferred for each time dependent variable of interest, a large undertaking.

Because the proportion of partial respondents relative to the entire sample was small in NMCES, an estimation strategy had advantages when it: (1) was straightforward in its derivation, (2) was directly applicable to a diverse set of time dependent analysis measures, (3) included partial respondents in analyses without sacrificing the quality of the final estimates, and (4) allowed the derivation of statistically valid estimates of utilization and expenditure measures expressed in terms of population means and totals. The data adjustment strategy to correct for partial data had these advantages and was easily implementable as an additional stage of weighting. For these reasons, the data adjustment strategy was selected as a viable approach for attrition compensation, recognizing the limitations encountered when attempting to generate person-level distributional estimates. For these estimates, the weight adjustment strategy should be employed.

The utilization and expenditure measures investigated in this study are not all inclusive and may not be completely sensitive to characteristics of partial respondents that would affect the estimates derived from the alternative attrition compensation strategies. Further studies, which consider additional health care indices, would be needed for a final determination of the effects of the alternative methods and the appropriateness of the adopted attrition compensation strategy.

References

Aday, L. A. and R. Eichhorn (1972). _The Utilization of Health Services: Indices and Correlates - A Research Bibliography_, DHEW Publication No. (HSM)73-3003, Washington DC: U.S. Government Printing Office.

Chapman, D. W. (1976). A Survey of Nonresponse Imputation Procedures. _Proceedings of the American Statistical Association, Social Statistics Section_, 245-251.

Cohen, S. B. (1982). An Analysis of Alternative Imputation Strategies for Individuals with Partial Data in the National Medical Care Expenditure Survey. Review of Public Data Use, 10, 153-165.

Ernst, L. R. (1978). Weighting to Adjust for Partial Nonresponse. Proceedings of the American Statistical Association, Survey Research Methods Section, 468-473.

Kalton, G., J. Lepkowski, and R. Santos (1981). Longitudinal Imputation: Income Survey Development Program, Report Available from the Survey Research Center, University of Michigan.

Kish, L. (1965). Survey Sampling, New York: John Wiley and Sons.

Phelps, C. E. and J. P. Newhouse (1974). Coinsurance and the Demand for Medical Services, Rand Report No. R96410EO/NC, Available from Rand Corporation, Santa Monica, California.

Shah, B. V. (1981). SESUDAAN: Standard Errors Program for Computing of Standardized Rates from Sample Survey Data, RTI Report No. RTI/5250/00-01S, Research Triangle Institute, Research Triangle Park, North Carolina.

Spiers, E. F. and J. J. Knott (1969). Computer Methods to Process Missing Income and Work Experience Information in the Current Population Survey. Proceedings of the American Statistical Association, Social Statistics Section, 289-297.

Woodruff, R. S. (1971). A Simple Method for Approximating the Variance of a Complicated Estimate. Journal of the American Statistical Association, 66, 411-414.

11
Imputation for Partial Nonresponse

Longitudinal surveys obtain data using multiple rounds of data collection. Consequently, sample attrition will result in partially missing data that must be dealt with prior to data analysis. The previous chapter discussed an investigation of alternative approaches to account for data missing due to survey attrition and described the reweighting approach used by the NMCES. This chapter presents the nonresponse and attrition rates encountered in the NMCUES and describes the implementation of an imputation procedure to replace attrition-related missing data.

Over the five rounds of data collection, attrition of initially responding individuals occurred at a rate of approximately five percent for the national household sample (HHS) and at a somewhat larger rate for the four State Medicaid Household Survey (SMHS) samples. Even this low a level of missing data could not be ignored, since many NMCUES analyses will use data aggregated over the full year. For example, estimating the distribution of 1980 medical care expenditures requires that complete annual data be available. In performing an analysis requiring annual data, two options are available. One could analyze only the complete data after reweighting the complete data records to compensate for the removal of partial data records. The second option is to perform the analysis using the full data set after imputing for the partially missing data.

The simplest solution to implement restricts analyses to individuals with complete data by reweighting of the complete data records to compensate for the removal of partial data records. This can be wasteful when many individuals have data that are almost complete. Further, since death and institutionalization result in high levels of survey attrition prior to the event, valuable information may be lost which cannot be adequately compensated for by sample weighting.

The second option is to perform analyses using the full data after imputing data for time periods for which data are missing. This approach has the disadvantage that nearly the entire year's data may be missing for some individuals and hence almost all their data will be imputed. There would seem to be little gain in including individuals with almost but not all their data missing. Further, imputation can have a greater variance inflation potential for survey estimates than weight adjustment procedures.

NMCUES decided that a combination of reweighting and imputation was the best approach. The two types of analysis units used in analyzing NMCUES data - individuals and families - required two sets of sample weights. Each weighting task defined the concept of respondent so that units were considered to be totally nonresponding when they provided data for less than one-third of the time in 1980 that they were eligible to provide data. The data that nonresponding units did provide are not to be included in analyses. The one-third of eligible time rule was based upon the perception that respondents with less data would not have provided enough information to allow their missing data to be accurately predicted. Chapter 7 documents the general procedure used to construct analysis weights that compensate for bias due to total nonresponse. This chapter describes the attrition in the NMCUES household surveys and the procedures that were used to replace the resultant missing data (Cox and Sweetland, 1982; Cox, 1983).

11.1 Level of Attrition-Related Missing Data

In order to replace data missing due to survey attrition, it was necessary to determine the dates for which each individual responded, the dates for which they were ineligible, and the dates for which they failed to respond. This information was used to determine whether or not each individual had provided data for all of the 1980 time period in which they were eligible for data collection.

The first step in this process was determining the response status of each sample individual for each round. Basically an individual's response status can be summarized as either responding, nonresponding, or ineligible. However, this response status could not be determined by examining the results of that round only, since an individual who failed to provide data in a round was allowed to retrospectively provide data for the time period in a future round. To facilitate the understanding of the true level of attrition for NMCUES, a modified version of a roundwise response status variable was used to calculate the level of attrition for each round of data collection after the first.

In Round 1, nonresponse was at the reporting unit (RU) level so the calculation of attrition rates was made for reporting units rather than individuals. These reporting units were families living within the same dwelling unit. When a family contained a college student living away from home, the student was assigned to a separate reporting unit of his own to facilitate data collection since the college student's interview had to be conducted independently from that of his family. Events such as marriage and divorce led to the formation of new families after Round 1 and hence the creation of new reporting units. Table 11-1 contains Round 1 response rates for reporting units in the national household component of the NMCUES and for the four SMHS sample components. This table was created from the control file used to monitor the day to day status of field operations, data receipt, and data entry. Only those reporting units contacted in Round 1 were utilized in this analysis. Thus, new reporting units that were created through subsequent divisions of initial Round 1 reporting units were not included in the analysis.

An examination of Table 11-1 reveals that 13.3 percent of the sampled housing units were ineligible for the National Household

Survey with the major reasons being vacancy or that the family occupying the unit had their usual residence elsewhere. An overall response rate of 91.1 percent was obtained from the eligible reporting units with the major reason for nonresponse being refusal to participate.

For the SMHS samples, Table 11-1 shows Texas with the highest rate of ineligible reporting units at 15.0 percent, and New York with 5.8 percent for the lowest rate. A marked difference in the overall Round 1 response rate can be observed between the four SMHS state samples with 96.9 percent for Texas, 87.3 percent for California, 82.6 percent for Michigan, and 79.8 percent for New York. Also of interest is the low refusal rate occurring in Texas with 0.9 percent of the eligible units refusing to respond. This rate is low compared to the 3.9 percent refusal rate encountered in California. The differences between the state results are likely due to differing program eligibility requirements.

For all rounds of data collection after Round 1, Table 11-2 contains response status results by round for individuals from initially responding RUs for each of the separate surveys. This analysis does not include those key individuals that entered the survey after Round 1 as newborns or returns from institutions or the armed forces. The roundwise response rates were consistently higher for the National Household Survey than for the Medicaid household surveys and the response rates over all the rounds were quite good. The difference between the general household survey and the Medicaid surveys is to be expected since the Medicaid surveys include propor-

Table 11-1. Reporting Unit Interview Final Results in Percentages for Round 1 of the NMCUES

Final Result	National Household Survey	State Medicaid Household Surveys			
		Texas	California	New York	Michigan
Total Sample	100.0	100.0	100.0	100.0	100.0
Ineligible Reporting Units	13.3	15.0	8.8	5.8	14.4
Eligible Reporting Units	86.7	85.0	91.2	94.2	85.6
Ineligible Reporting Units	100.0	100.0	100.0	100.0	100.0
Vacant Housing Unit	49.3	0.0	0.0	0.0	0.0
Demolished Housing Unit	12.2	0.0	0.0	0.0	0.0
Merged Housing Unit	2.3	0.0	0.0	0.0	0.0
Not a Housing Unit	13.5	0.0	0.0	0.0	0.0
Vacation/Second Home	16.1	0.0	0.0	0.0	0.0
Entire RU in Armed Forces	1.5	0.0	0.0	0.0	0.0

Entire RU Ineligible Students	1.4	0.0	0.0	0.0	0.0
Entire RU Institutionalized	0.8	89.8	82.8	51.6	88.4
Entire RU Deceased	0.7	9.2	14.1	20.9	9.1
Other Out of Sample	2.2	1.0	3.1	27.5	2.5
Eligible Reporting Units	100.0	100.0	100.0	100.0	100.0
Questionnaire Complete	91.1	96.9	87.3	79.8	82.6
Questionnaire Not Complete	8.9	3.1	12.7	20.2	17.4
No One Found at Home	1.0	0.6	2.3	3.9	2.4
Entire RU Case Moved	0.1	1.0	5.4	10.3	9.3
Language Barrier	0.0+	0.0	0.0+	0.1	0.0
Physically/Mentally Incompetent	0.2	0.0+	0.3	0.8	0.1
Refusal	7.2	0.9	3.9	3.8	3.5
Breakoff	0.1	0.0+	0.4	0.4	0.2
Other In Sample	0.3	0.4	0.3	0.9	1.7
Percentage Base					
Total Sample	8,359	1,440	1,449	1,560	1,683
Ineligible Reporting Units	1,115	216	128	91	242
Eligible Reporting Units	7,244	1,224	1,321	1,469	1,441

Table 11-2. Interview Final Results as a Percentage for Rounds 2-5 for Key Individuals From Initially Responding RUs

Sample	Round	Final Result			Base o
		Data Obtained	Data Not Obtained	Ineligible	Percenta
National Household Survey	2	99.5	0.4	0.1	16,90
	3	97.9	1.5	0.6	16,90
	4	97.1	2.0	0.9	16,90
	5	96.5	2.3	1.2	16,90
SMHS Texas	2	99.3	0.3	0.4	3,90
	3	96.8	2.3	0.9	3,90
	4	94.4	4.6	1.0	3,90
	5	93.8	4.6	1.6	3,90
SMHS California	2	98.6	1.0	0.4	3,01
	3	94.6	4.1	1.3	3,01
	4	92.8	5.4	1.8	3,01
	5	91.6	6.1	2.3	3,01
SMHS New York	2	98.8	1.0	0.2	2,92
	3	96.5	2.3	1.2	2,92
	4	94.8	3.5	1.7	2,92
	5	94.2	3.6	2.2	2,92
SMHS Michigan	2	99.2	0.4	0.4	2,78
	3	96.3	2.7	1.0	2,78
	4	94.8	4.3	0.9	2,78
	5	93.9	4.7	1.4	2,78

tionally more aged, disabled, and economically disadvantaged persons.

The roundwise response status indicators were used to determine when an individual provided complete annual data. The determination of whether complete annual data were available for sample individuals was restricted to individuals who had belonged to a household that responded in Round 1. These complete data rates will be presented after the imputation strategy is discussed.

11.2 The Attrition Imputation Procedure

When an individual fails to provide data for one or more data collection rounds, health care utilization will be missing for the time period covered by these rounds. Ignoring this missing data would obviously result in underestimation of utilization and expenditures since an accurate count of the number of visits or total charges for 1980 cannot be obtained for these individuals. The NMCUES solution to this problem was to replace the missing data by data provided by a sample individual with complete annual data.

Each individual with incomplete annual data, referred to later as "nonrespondents" or "recipients," was linked to an individual with complete annual data, referred to later as "respondents" or "donors." For time periods for which the recipient had data missing, the visits reported by the donor for that time period, if any, were imputed to the recipient. If the donor was ineligible to provide data during the time period for which the recipient had data missing, the recipient was imputed to be ineligible during that time period also. The reasoning underlying the imputation procedure was that when data were not collected for an individual, not only is the medical care experience unknown but whether or not the individual was eligible to provide data is also unknown.

The nature of the attrition imputation task required that some form of hot deck imputation be used to replace the missing data. For this task, the weighted sequential form of hot deck imputation was deemed to be most appropriate because of the additional control that it could provide over the imputation procedure. The first step was to sort the two data files with respect to variables related to response and the distribution of characteristics of interest. Both files had the nonresponse-adjusted reporting unit weight attached to each record. The imputation occurred within imputation classes defined in such a manner that within each class nonrespondents and respondents should have similar responses. Use of weighted sequential hot deck insured that the distribution of weighted means and proportions would be preserved within each class over repeated imputations. The selection of classing and sorting variables is discussed in the next section.

An intermediate product at this stage was a nonrespondent data file with the dates for which data were missing and the Participant Identification Number (PID) of the respondent who was linked to the nonrespondent via the weighted sequential hot deck algorithm. The use of this data file to create imputed visit records is described after discussing the process used to select classing and sorting variables.

11.3 Determination of Classing and Sorting Variables

As discussed in Chapter 8, the weighted sequential hot deck imputation procedure is based upon a model for the nature of data missingness. The required model for this procedure specifies that respondents and nonrespondents can be partitioned into classes in such a way that within classes either the data are missing at random or the data that are missing can be explained in terms of a distance function associated with some quantitative variable. The distance function concept implies that if data records within classes are sorted by the quantitative variable, then the records close together in the data file will have similar responses.

NMCUES sample data were analyzed to obtain insight into the variables that should be used to model the presence or absence of complete annual data. The following data were available for use in classifying individual records or for sorting within classes:

- age of individual,
- race of individual,
- sex of individual,
- years of school completed by household head,
- number of ambulatory medical visits in the first quarter of 1980,
- health status of individual,
- health insurance coverage, and
- State Medicaid Aid category.

In forming imputation classes, the overall goal was to form classes for which responses were homogeneous within the classes and heterogeneous between classes and for which the rate of missing data varied. Further, the characteristics used to define the classes had to be known for both respondents and nonrespondents. This necessarily implied that the level of missing data for classing variables had to be low.

In addition to determining the classing variables that correlated best with health care response variables and with the level of incomplete data, missing data for the classing variables had to be assessed as well. Classing variables with a high level of missing data were rejected for use, especially when missing data for the classing variable was linked to the presence of incomplete annual data. Other variables with moderate to low levels of nonresponse had their missing data imputed when they were selected for use in forming imputation classes.

In order to form efficient imputation classes, the proportion of individuals providing complete annual data was calculated for the domains defined by each level of the potential classing variables. These annual completion rates are presented in Table 11-3. Complete annual data were obtained from 94 percent of the persons initially responding in the National Household Survey and for 85.6 to 92.3 percent in the State Medicaid Household Surveys. The variables most related to the level of complete data appeared to be education of the household head, health status, age, and health insurance coverage.

Note that the annual completion rate for each survey is lower than the response status rates, which were presented earlier, would suggest. The earlier rates were based upon round by round results. Frequently, individuals failed to provide data in a round but were not asked to provide data for the missing time period in later rounds. This resulted from the fact that completed interviews were defined on an reporting unit rather than a person level. In addition, the attrition rates for Round 5 classified all individuals who became ineligible prior to Round 5 as "ineligible" in Round 5, which automatically implies all relevant data were obtained for the Round 5 interview. However, data were frequently missing for the time period directly prior to when they became ineligible through death, institutionalization, leaving the country, or joining the armed forces. This factor also contributes to the differences between the two rates.

Estimates of the average number of annual ambulatory medical care visits were also made for the domains defined by the classing variables. These estimates were obtained from individuals who had complete annual data (Table 11-4). Certain variables can be seen to have levels that are highly correlated with the annual number of medical visits: age of individual, sex of individual, number of medical visits in first quarter, health insurance coverage, and for the SMHS samples, State Medicaid Aid category.

Although age appeared to be an obvious classing variable, it was not used in classing for the SMHS states due to small class sizes. A larger total sample size for HHS allowed age to be used as a classing variable in the National Household Survey imputation. The sex variable was selected as a within classes sorting variable for each of the five sample types. Differences in annual visit counts were observed between the sexes, but differences in completion rates were not sufficiently large to suggest its use as a classing variable. For all five sample types, the variables for health insurance coverage and number of medical visits in the first quarter showed strong differences in average annual visit count between levels. Equally as important, these variables also showed strong differences in completion rates between levels. Within the SMHS States, the Medicaid Aid Category variable also exhibited differences in both annual visit counts and completion rates between levels. This evi-

Table 11-3. Annual Complete Data Rates for Key Individuals from Responding Round 1 RUs

Variable	Level	Sample Type				
		HHS	SMHS California	SMHS Michigan	SMHS New York	SMHS Texas
Total		94.0	85.6	90.3	89.2	92.3
Age of Individual	0-16	94.7	86.1	90.4	92.2	91.6
	17-29	93.1	81.4	89.6	87.0	88.4
	30-44	94.7	84.7	89.8	92.0	91.3
	45-54	94.4	85.5	91.5	85.1	93.1
	55-64	94.5	90.1	88.5	86.4	95.8
	64+	92.7	87.4	91.5	88.0	96.2
Race of Individual	Black	92.2	81.3	85.5	87.7	90.5
	Nonblack	94.3	86.5	93.5	89.9	93.2
Sex of Individual	Male	93.9	84.5	89.9	88.3	91.6
	Female	94.1	86.4	90.6	89.7	92.8
Education of Head of Household in Years	0	82.7	96.5	95.6	89.2	97.1
	1-8	94.4	83.3	89.7	89.8	93.9
	9-12	93.9	86.4	90.1	89.3	90.1
	13+	94.7	85.0	92.5	87.0	83.6

Number of Medical Visits in First Quarter	0	93.2	84.6	89.2	88.0	92.0
	1	95.1	84.0	93.2	89.0	93.5
	2	94.4	87.9	90.1	89.0	92.2
	3-4	95.6	88.4	91.9	93.2	93.2
	5-6	95.9	85.4	91.9	92.4	87.6
	7-8	94.4	92.3	91.3	85.7	92.4
	9+	93.6	84.9	83.6	87.6	93.8
Health Status	Excellent	95.1	86.5	91.1	90.7	90.6
	Good	93.9	86.5	92.4	91.0	93.4
	Fair	92.9	87.5	88.4	87.2	92.9
	Poor	91.0	84.0	86.6	87.7	93.8
Health Insurance Coverage	Medicare	92.9	85.8	91.4	89.5	93.4
	Other Public	92.9	81.0	83.9	80.0	93.6
	Private	95.3	87.0	91.0	91.1	90.6
	Uninsured	87.2	81.5	81.5	82.2	89.6
State Medicaid Aid Category	SSI Aged	N.A.	85.2	91.2	87.7	91.6
	SSI Blind /Disabled	N.A.	87.1	88.5	89.3	95.9
	AFDC	N.A.	85.2	91.5	91.8	91.2
	State Only	N.A.	85.2	87.5	86.3	N.A.

Table 11-4. Average Annual Number of Medical Visits for the Domains Defined by the Levels of the Potential Classing Variables

		Sample Type				
Variable	Level	HHS	SMHS California	SMHS Michigan	SMHS New York	SMHS Texas
Total		5.11	7.29	6.82	9.16	4.70
Age of Individual	0-16	3.73	3.54	3.88	4.56	2.27
	17-29	4.43	4.83	5.38	8.03	3.60
	30-44	4.92	7.48	8.43	14.96	5.60
	45-54	5.54	10.03	10.10	12.36	6.69
	55-64	6.48	11.44	8.93	13.23	8.17
	65+	7.96	11.83	9.75	13.78	7.77
Race of Individual	Black	3.90	7.73	5.25	7.56	3.84
	Non-Black	5.26	7.20	7.80	9.91	5.13
Sex of Individual	Male	4.32	5.34	5.75	8.52	3.43
	Female	5.83	8.74	7.64	9.60	5.56
Education of Head of Household in Years	0	6.57	8.23	4.92	8.91	5.54
	1-8	5.22	7.79	7.32	10.44	4.76
	9-12	4.83	6.85	6.21	8.01	4.26
	13-18	5.42	7.63	9.37	10.78	5.23

Number of Medical Visits in First Quarter	0	1.98	2.09	2.09	2.30	1.62
	1	4.18	4.23	4.57	4.77	3.76
	2	6.53	7.08	7.08	7.04	6.46
	3-4	9.53	11.39	10.64	10.75	10.20
	5-6	14.56	16.10	15.82	17.27	14.22
	7-8	19.02	22.00	21.17	23.14	19.06
	9+	38.45	44.62	40.43	70.92	40.32
Health Status	Excellent	3.64	3.70	3.48	3.51	2.14
	Good	5.27	5.42	5.51	6.37	2.86
	Fair	9.05	9.66	10.47	13.10	6.84
	Poor	13.54	19.55	13.83	23.40	12.31
Health Insurance Coverage	Medicare	8.75	8.32	7.88	10.60	5.84
	Other Public	5.45	5.86	7.84	8.82	6.37
	Private Plans	4.59	4.32	3.66	3.56	2.67
	Uninsured	2.54	2.62	2.80	3.73	1.86
State Medicaid Aid Category	SSI Aged	N.A.	9.61	8.90	12.85	5.66
	SSI Blind/ Disabled	N.A.	10.02	8.86	10.80	6.66
	AFDC	N.A.	5.14	5.08	5.46	3.15
	State Only	N.A.	6.33	6.28	10.35	N.A.

dence led to the use of the variables for health plan coverage and first quarter medical visits to form imputation classes for all five of the sample types. Likewise, Medicaid Aid Category type was used for imputation class construction in each of the four SMHS samples.

The criteria for the selection of classing and sorting variables were generally consistent for the five sample types. Certain variables were shown to be important imputation classing variables regardless of sample type. To summarize the selection results, the National Household Survey imputation used age of individual, health plan coverage, and medical visits in first quarter as classing variables with sex of individual as a within class sorting variable. Within each of the four SMHS sample imputations, the variables chosen were the same: Medicaid Aid Category, health plan coverage, and medical visits in first quarter as classing variables with sex of individual as a within class sorting variable.

11.4 Imputation of Visit-Level Records

Unlike item-level imputation where particular item responses are created, the attrition imputation procedure created complete annual sets of event-level data records for the recipient person based upon the event-level records of the donor person. Health provider visits, health service charges, and other health care experiences of the donor person were copied to create new event-level records for the recipient's time periods of missing data. The data files involved in the imputation were those associated with medical provider visits, dental visits, hosital stays, prescribed medicines, and other medical expenses.

The imputation software program generated a nonrespondent data file containing the dates for which data were missing and the Participant Identification Number (PID) of the respondent who was linked to the nonrespondent for the purpose of supplying missing data. The time period for which the nonrespondent had missing data was then divided into two categories: imputed eligible days and imputed ineligible days. Imputed eligible days were those days for which the respondent donor was eligible for the survey and hence records were available to impute to the nonrespondent recipient. If the respondent donor was ineligible during a part of the recipient's missing data period, the recipient was classified as ineligible for the same time period. For the recipient's imputed eligible days, health care event records were supplied from the donor based upon the date associated with the event. The actual imputation of data records occurred for the event-level files. The result of the imputations for the event files is summarized in Table 11-5.

Some items contained in the donor's data records would not be meaningful when imputed to the recipient. Responses pertaining to the unique identification of the donor were not used; instead the recipient's values were used in the imputed record. In addition, response items concerning disability days and non-prescription medicines were not included in attrition imputation.

The results of attrition imputation were also used in weight development to define responding persons and responding families. The person-level weighting tasks defined individuals to be responding when they provided data for one-third or more of the time in which they were known or imputed to be eligible. The HHS family

Table 11-5. Results of Attrition Imputation for the Individual Event Files (HHS and SMHS Combined)

	File Type			
	Dental Visits	Hospital Stays	Medical Visits	Medical Expenses
Record Count Before Imputation	33,251	7,026	179,713	116,928
Imputed Records	2,452	430	5,673	4,252
Donor Records Not Used	1,869	471	16,300	9,528
Donor Records Used Once	2,352	422	5,491	4,123
Donor Records Used Twice	50	4	91	63
Donor Records Used Three Or More Times	0	0	0	1
Total Donor Records Available	4,271	897	22,882	13,715
Record Count After Imputation	35,703	7,456	185,386	121,180

weighting task defined families to be responding when they contained at least one person defined to be responding by the previous rule and when data were available for one-third of the time that the family was known or imputed to be eligible to provide data. The SMHS family weighting tasks defined families to be responding when at least one Medicaid casemember responded and when data were available for one-half of the time the family casemembers were known or imputed to be eligible.

At the time that the attrition imputation was implemented, variables defining overall response status and periods of family membership were not available. Instead, all persons, key and nonkey, who were linked to ever-responding Round 1 households were included in the attrition imputation task. Thus, data were imputed for nonkey individuals and individuals who were subsequently defined to be nonresponding. Hence, many of the imputed visit records shown in Table 11-5 will not be used in person-level analyses. Similarly, data were imputed for individuals for time periods when they were not members of responding key families. Many of the imputed data records shown in Table 11-5 will also not be used for family-level analyses.

11.5 Use of Attrition-Imputed Data

The approach that NMCUES chose for dealing with missing data due to survey attrition is a cost effective approach. Since imputation can increase the variance of survey estimates, discarding the data for individuals who respond for less than one-third of the time in which they were eligible and reweighting the data for respondents should produce estimates with a smaller mean square error. However, reweighting itself introduces variability and may not totally remove the bias due to nonresponse. Hence, it makes sense to include individuals who provide almost-complete data and to complete their data by imputation.

Each visit-level record in the data base received an imputation indicator which designates whether the data are imputed or real. The analyst using NMCUES data must carefully consider whether attrition-imputed records should be included in his analyses. When individual-level estimates of utilization and expenditures are being constructed, attrition-imputed visits should be included. Visit-level analyses will usually include the imputed visits. However, for certain analyses, it may be misleading to include the imputed visits. If one wishes to form estimates of the incidence of cancer, for instance, by assessing the number of individuals reporting visits for cancer treatment, then the imputed visits should not be used to classify the individual. Since more often than not a visit from a donor with cancer will have been imputed to a recipient without cancer, using both imputed and real visit records to classify individuals will result in overestimating the number of individuals with cancer. In general, data from attrition-imputed records should not be used when the purpose is to classify the individual according to presence or absence of a particular type of visit record. The imputed data will generally be used when individual-level or family-level aggregates are required and for most visit-level analyses.

When imputed data are used, the estimates will contain additional variability due to imputation. There is no readily available method to estimate the variance of the resultant statistics. However, since the number of imputed records is small, the increase in variance should be small as well. Since the variance for NMCUES estimates will usually be calculated as the sum over strata of the within stratum squared difference between the two replicate linearized estimates, it is reasonable to presume that the use of attrition-imputed data should not contribute greatly to the differences between the two replicate estimates and hence, the variance induced by imputation should be negligible for most analyses.

References

Cox, B. G. (1983). Compensating for Missing Data in the NMCUES. Priorities in Health Statistics 1983, DHHS Publication No. (PHS) 81-1214, National Center for Health Statistics, Hyattsville, Maryland, 186-191.

Cox, B. G., and S. S. Sweetland (1982). Imputation of Attrition-Related Missing Data for the National Medical Care Utilization and Expenditure Survey, RTI Report No. RTI/1898/06-01F, Contract No. HRA-233-79-2032, Health Care Financing Administration, Baltimore, Maryland and the National Center for Health Statistics, Hyattsville, Maryland.

12

Statistical Analysis of Complex Survey Data

As surveys are conducted with varying design considerations, analytical concerns, and populations of interest, it has become common to depart from equal probability of selection methods. Disproportionate sampling is often adopted to insure sufficient representation of small subgroups, while simultaneously yielding reliable estimates of characteristics of the entire population. Even when sample members are selected with equal probability, the need to obtain the greatest precision for fixed cost and time constraints almost always leads to complex sampling procedures rather than the use of simple random sampling. Stratification and proportionate representation are two widely used design features to increase precision. Area sampling is a third method used in combination with stratification and probability selection schemes to reduce survey costs. These selection procedures are described in Chapter 2.

Even when simple random sampling is used, the unweighted data will not produce unbiased estimates due to nonresponse and coverage deficiencies. Generally, specific subgroups of an underlying population are more prone to refuse participation in the survey, forcing the need to consider nonresponse adjustments to minimize bias when estimating relevant population parameters and totals. In addition, some population members may not be included in the frame and hence not have the proper opportunity to be included in the sample.

These methods of sampling are components of the set of survey strategies which create a complex survey design. These complexities must be considered in the estimation process through the development of analysis weights, which reflect unequal selection probabilities and include post-stratification or weighting class adjustments to account for survey nonresponse. In this setting, unbiased national estimation of population parameters requires the use of these analysis weights with survey data. The development of analysis weights for survey data is described in Chapter 7.

Due to the departure from the simple random sampling assumptions which form the basis for most statistical procedures, survey data require special procedures for variance estimation and analysis. This chapter presents methods of variance estimation that are appropriate for complex survey data, with a focus on those strategies suitable for surveys such as the NMCES. The discussion includes an identification of available statistical software packages that generate variance estimates of descriptive statistics such as population means, totals, and ratios. Analysis methods are also presented that accommodate survey design complexities.

12.1 Variance Estimation

When complex survey designs are used, the conventional methods of variance estimation, construction of confidence intervals, and hypothesis testing are not appropriate due to the departure from simple random sampling. For most survey designs, the assumptions of independence and equal probability of selection inherent in simple random sampling designs will no longer be true. Sample units are not independent due to the correlation induced by cluster sampling and several stages of stratification.

Standard methods of variance estimation for means, proportions, ratio estimates and regression coefficients, which are present in the most commonly used statistical package programs, assume simple random sampling (SRS). Examples of such software packages include SPSS (Statistical Package for the Social Sciences, Francis, 1983), SAS (Statistical Analysis System, Helwig and Council, 1979) and BMDP (Biomedical Data Program, Francis and Berk, 1978). When SRS approaches are used for data from a complex survey design, the true variance is generally underestimated. The consequences of using variance estimates derived in this manner in the construction of confidence intervals and for statistical inference result in anti-conservative tests, where the probability of rejecting the null hypothesis is greater than specified by the α-level of the test.

Several methods to approximate the sampling variance have been developed that incorporate the components of complex survey designs in their derivations. The three most generally accepted and frequently used techniques are balanced repeated replication, the jack-knife method, and the Taylor Series linearization method. Each of these methods of variance estimation assumes a multi-stage, stratified sampling design with at least two primary sampling units drawn from each stratum. When the requirement of two selections per stratum is not satisfied, the primary sampling units must be paired to form pseudostrata with two PSUs per pseduostratum for variance computation. This procedure was needed for the NMCES where deep stratification resulted in one selection per stratum.

12.1.1 Balanced Repeated Replication

The replication approach to variance estimation was developed at the Bureau of the Census (Gurney, 1969) from basic replication concepts of sampling (Mahalanobis, 1939 and Deming, 1956). Several of these strategies suggested that a solution to the problem of obtaining variance estimates for survey statistics could be found by interweaving a number of sample replicates within the sample design. After sample selection, replications can be obtained by reproducing the complex design of the complete sample, which in the NMCES entails random selection of one of the two primary sampling units from each stratum. The sample weights are adjusted to produce national estimates of desired parameters from each half of the parent sample. An estimate of the desired population parameter, μ, is computed from the half-sample. The squared deviation of the half-sample estimate, $\hat{\mu}_H$, from the full-sample estimate, $\hat{\mu}_F$, provides an estimate of the variance of the full sample statistic. That is,

$$\mathrm{Var}(\hat{\mu}_F) = (\hat{\mu}_H - \hat{\mu}_F)^2 \tag{12-1}$$

Since each of these variance estimates are highly variable, replication is used to produce a more stable estimate. The most widely used variance estimator of this type is expressed as the average over the K replications of the squared deviation of the half sample estimate from the full sample estimate or

$$\mathrm{Var}(\hat{\mu}_F) = \sum_{i=1}^{K} (\hat{\mu}_{Hi} - \hat{\mu}_F)^2 / K \quad (12\text{-}2)$$

where $\hat{\mu}_{Hi}$ is the mean for the i-th half-sample and K is the number of independent half sample selections. Note that the complement of each half sample also produces a variance estimate, or

$$\mathrm{Var}(\hat{\mu}_F) = \sum_{i=1}^{K} (\hat{\mu}_{Ci} - \hat{\mu}_F)^2 / K \quad (12\text{-}3)$$

where $\hat{\mu}_{Ci}$ is the parameter estimate computed from each of the corresponding complement half-samples. A more reliable variance estimate is obtained by combining the two estimators or

$$\mathrm{Var}(\hat{\mu}_F) = \{ \sum_{i=1}^{K} (\hat{\mu}_{Hi} - \hat{\mu}_F)^2 + \sum_{i=1}^{K} (\hat{\mu}_{Ci} - \hat{\mu}_F)^2 \} / 2K \quad (12\text{-}4)$$

An alternative estimator that is equivalent for linear statistics is

$$\mathrm{Var}(\hat{\mu}_F) = \sum_{i=1}^{K} (\hat{\mu}_{Hi} - \hat{\mu}_{Ci})^2 / 4\,K \quad (12\text{-}5)$$

The precision of these variance estimators is known to increase with the number of replications, but the maximum precision is achieved only gradually when no systematic pattern of selection is considered. Full precision is obtained from a sample design with H strata by generating the $K = 2^H$ possible replications. However,

McCarthy (1966) demonstrates that an orthogonally balanced set of half-samples can produce a variance estimator with the same precision as if all possible replications were included. This procedure, referred to as balanced repeated replication, results in a substantial reduction in the required number of half sample replications and in computer time and cost.

Using balanced repeated replication, an orthogonally balanced set of half-samples is obtained by initially determining the number of required replications. This value, K, is equal to the number of sample strata, H, when integrally divisible by 4 or by the next greatest integer that satisfies this constraint. Primary sampling units within strata are designated by a = 1,2. A K by H matrix is then formed with entries, X_{kh}, consisting of +1 or -1 and columns (strata) that are orthogonal to each other. Here,

$$\sum_{k=1}^{K} X_{kh} X_{kh'} = 0 \qquad (12\text{-}6)$$

where $h \neq h'$. Half-sample selection is determined by the matrix patterns. Primary sampling units designated a = 1 are included in the half sample replicate when $X_{kh} = +1$; similarly, primary sampling units designated a=2 are included when $X_{kh} = -1$. For NMCES, the number of replications required to produce an orthogonally balanced set is 72; the design matrix pattern is summarized in Table 12-1.

The construction of these orthogonally-balanced matrices is described by Plackett and Burman (1946). Bean (1975), Frankel (1971), and Kish and Frankel (1974) describe empirical investiga-

Table 12-1. Balanced Repeated Replication Design Matrix

Half-Sample Replicate	Strata							
	1	2	3	4	.	.	68	69
1	+1	-1	-1	-1	.	.	+1	+1
2	+1	+1	-1	-1	.	.	+1	+1
3	+1	+1	+1	-1	.	.	+1	+1
.	.	.	.	.	.	.	.	.
.	.	.	.	.	.	.	.	.
70	-1	-1	-1	-1	.	.	+1	+1
71	-1	-1	-1	-1	.	.	+1	+1
72	-1	-1	-1	-1	.	.	-1	-1

tions in which variance estimates derived by balanced repeated replication are contrasted with Taylor Series and jack-knife estimates.

12.1.2 Jack-Knife Method

The application of the jack-knife method to variance estimation was first suggested by Tukey (1958) and its use extended by Brillinger and Tukey (1964) and Brillinger (1964, 1966). Jones (1974) provides a theoretical justification for the use of the jack-knife in a sampling context. Jack-knife variance estimates measure the contribution to the variance for each of the strata by eliminating replicates from the sample. Specifically, let $\hat{\mu}_h$ represent the parameter estimate generated from the sample when one primary sample unit in the h-th stratum is eliminated and replaced by the remaining primary sampling unit in the stratum. Data from the remaining primary sampling unit are then included twice in the estimation procedure. Similarly, let $\hat{\mu}_{hc}$ represent the estimate the complement replicate when the primary sampling units are interchanged. Once H replicates have been formed, one for each stratum, variances can be estimated by one of the two following procedures:

$$\text{Var}(\hat{\mu}) = \left[\sum_{h=1}^{H} (\hat{\mu}_h - \hat{\mu})^2 + \sum_{h=1}^{H} (\hat{\mu}_{hc} - \hat{\mu})^2 \right] / 2 \quad (12\text{-}7)$$

or

$$\text{Var}(\hat{\mu}) = \sum_{h=1}^{H} (\hat{\mu}_h - \hat{\mu}_{hc})^2 / 4 \quad (12\text{-}8)$$

where $\hat{\mu}$ is the full sample parameter estimate (Kish and Frankel, 1974).

12.1.3 Taylor Series Linearization Method

The use of a Taylor Series linearization to obtain variance estimates for ratio statistics is due to Keyfitz (1957). The *Taylor Series linearization method* yields approximate variance estimates of first-order statistics, based upon the variance of the linear terms of the Taylor expansion for the respective statistics (Kish and Frankel, 1974; Brillinger and Tukey, 1964). Tepping (1968) described the Taylor Series expansion for regression coefficients; the linearized form was developed by Folsom (1974) and Fuller (1975).

Generally, the complex statistics that are considered in NMCES (e.g., ratio estimators, regression coefficients) can be expressed as a function of m estimated totals which are derived from the sample; that is,

$$Y = f(Y_1, Y_2, \ldots, Y_m) \tag{12-9}$$

The large-sample approximation to the variance of a complex sample estimator, Y, takes the form

$$E \left\{ \sum_{i=1}^{m} \frac{\partial Y}{\partial Y_i} [Y_i - E(Y_i)] \right\}^2 \tag{12-10}$$

where the partial derivatives, $\partial Y/\partial Y_i$, are evaluated at expected values (Woodruff, 1971). This expression considers the first-order Taylor Series approximation of the deviation of Y from its expected value.

12.1.4 Available Statistical Software for Variance Estimation

The variance estimation strategies have been incorporated as procedures in several of the widely used statistical packages. The balanced half-sample replication method of variance estimation for means, proportions, totals and rates can be implemented through the Health Examination Survey Variance and Cross Tabulation Program developed by the National Center for Health Statistics (Jones, 1977). Appropriate variance estimates for means, proportions, and totals using the Taylor Series linearization method can be generated through the SESUDAAN program - Standard Errors Program for Computing of Standardized Rates from Sample Survey Data - which is accessible through SAS (Shah, 1979). The same capacity exists in the OSIRIS IV, Statistical Analysis and Data Management Software Systems Package, through application of the PSALMS procedure (Van Eck, 1979). Similarly, variance estimates of ratios can be generated through the SAS accessible procedure, RATIOEST: Standard Errors Program for Computing of Ratio Estimates from Sample Survey Data, also using a linearization approximation (Shah, 1981).

OSIRIS IV provides access to repeated replication procedures to compute estimates of regression coefficients and their sampling errors for data from complex survey designs. The OSIRIS IV Repeated Replication Sampling Error Analysis Procedure, REPERR, allows for the creation of replications using one of three methods: balanced half-sample, jack-knife, or user specified replications. The complementary package to generate standard errors of regression coefficients by a Taylor Series approximation using complex survey data is the SURREGR program: Standard Errors of Regression Coefficients from Sample Survey Data (Holt, 1977).

12.2 Statistical Analysis of Complex Survey Data

Once an approximately unbiased variance estimate has been obtained for a particular statistic of interest, it is used in the construction of confidence intervals or incorporated into relevant test statistics to facilitate hypothesis testing. For complex survey designs, care must be taken in the specification of an appropriate test statistic for research hypotheses of interest. In the NMCES, the most frequently considered statistical analyses can be separated into six distinct classes. These distinct types of statistical analyses are distinguished as follows:

- Univariate tests of hypotheses about the value of specific population parameters. This class is most typically characterized by tests of a population mean, proportion, ratio or total.
- Hypotheses concering the difference between specific statistics (e.g., means, proportions, ratios, totals) that characterize two subgroups or domains of the total population.
- Multivariate analyses that assess the relationship between several predictor variables and a single continuous dependent variable. Multiple regression models are used to test research hypotheses.
- Hypothesis testing within an analysis of variance framework. A set of factors, each with several levels, is examined to determine the extent to which they explain the variability of a criterion variable.
- Contingency table analyses that define criterion variables in terms of categorical response profiles and then classify respondents in terms of the response profile and potentially related independent factors expressed in categorical form. Inferences are made regarding the relationship between the factors and response variable.
- Multivariate analyses that assess the relationship between several predictor variables and a single binary dependent variable. A logistic regression is used to test relevant hypotheses.

In this section, a detailed description is presented of these distinct types of statistical analyses of complex survey data.

12.2.1 Hypothesis Testing for a Single Population Parameter

A major analytical concern of many surveys is the unbiased estimation of parameters that characterize the nation. These estimates are generally expressed in terms of population means, proportions, ratios and totals. Although the sample estimate is unbiased in expectation, its value for a particular sample will not necessarily be equal to the true population value, due to sampling variation. Consequently, population estimates are often presented in terms of confidence intervals. Depending upon the confidence level specified, a range is developed for the population parameter based upon the survey estimate.

To illustrate, a population mean, $\bar{Y}$, is estimated in the following manner:

$$\bar{y} = \sum_{i \varepsilon S} W(i)\, Y(i) \Big/ \sum_{i \varepsilon S} W(i) \qquad (12\text{-}11)$$

where Y(i) is the value of the criterion variable for the i-th member of the sample and W(i) is the i-th participant's analysis weight. A ratio estimator is used, to incorporate the weighting scheme which characterizes the complex survey design and yields an approximately unbiased mean estimate. Further, let $SE(\bar{y})$ represent the estimated standard error of the population estimate, which has incorporated the complexities of the survey design in its derivation. Define $1-\alpha$ as the confidence coefficient desired for the interval estimate. Since the ratio estimator follows an asymptotic normal distribution by the central limit theorem, the interval $\bar{y} \pm Z(1-\alpha/2)\, SE(\bar{y})$ represents a $100(1-\alpha)$ percent confidence interval for $\bar{y}$. In this setting, $Z(1-\alpha/2)$ represents the standardized normal variate at the $100(1-\alpha/2)$ percentile. Depending upon the value of α, this interval contains the population mean with $100(1-\alpha)$ percent confidence. Confidence intervals for parameters expressed in terms of proportions, ratios and totals are derived similarly.

Often, it is of interest to directly test concerning the values of specific population parameters. Relevant hypotheses include:

I. H_o: $T = T_o$

H_a: $T = T_a$

which considers a simple null hypothesis versus a simple alternative,

II. H_o: $T \leq T_o$

H_a: $T > T_a$

which considers a composite one-sided null hypothesis versus a one-sided composite alternative, and

III. H_o: $T \geq T_o$

H_a: $T < T_a$

which considers a composite one-sided null hypothesis versus a one-sided composite alternative. In these tests, T represents the population parameter of interest, expressed in terms of a mean, proportion, ratio, or total.

The following is the general formula for the appropriate test statistic:

$$Z = (\hat{T} - T_o) \ / \ SE(\hat{T}) \qquad (12\text{-}12)$$

where $\hat{T}$ is the sample estimate of the population parameter, weighted appropriately to account for the differential sample selection probabilities, nonresponse and post-stratification adjustments, and $SE(\hat{T})$ is the standard error of the estimate, adjusted for survey design complexities. The statistic has an asymptotic Gaussian form, according to the central limit theorem. For testing hypotheses of form I, a two-sided critical region is specified at the α level, whereas, a one-sided rejection region applies for hypotheses of form II and III. Normal Z-tests apply to test research hypotheses.

12.2.2 Hypothesis Testing for Two Population Subgroups

Many research hypotheses involve comparisons of parameters across two distinct population subgroups or domains. As noted, parameters of interest are generally expressed in terms of means, proportions, ratios and totals. Relevant hypotheses include:

I. $H_o: T_1 - T_2 = C$

$H_a: T_1 - T_2 \neq C$

which tests to determine if the difference is equal to C or unequal to C,

II. $H_o: T_1 - T_2 \leq C$

$H_a: T_1 - T_2 > C$

which tests to determine if the difference is less than or equal to C or greater than C, and

III. $H_o: T_1 - T_2 \geq C$

$H_a: T_1 - T_2 < C$

which tests to determine if the difference is greater than or equal to C versus less than C. T_i represents the population parameter of interest for subgroup i and C is a specified constant. Most often, C is set at zero to test equivalence of parameters across subgroups.

The following is the general formula for the appropriate test statistic:

$$Z = [(\hat{T}_1-\hat{T}_2)-C] \;/\; SE(\hat{T}_1 - \hat{T}_2) \tag{12-13}$$

where $\hat{T}_i$ is an approximately unbiased estimate, appropriately weighted, of the population parameter under consideration for subgroup i, and SE $(\hat{T}_1 - \hat{T}_2)$ is the standard error of the estimated difference. The standard error is expressed as:

$$SE(\hat{T}_1-\hat{T}_2) = [Var\ (\hat{T}_1) + Var(\hat{T}_2) - 2Cov(\hat{T}_1,\hat{T}_2)]^{\frac{1}{2}} \tag{12-14}$$

and estimated by replication, jack-knife or Taylor Series linearization procedures to account for the complex survey design. This complex statistic has an asymptotic Gaussian form, again according to the central limit theorem. Consequently, normal Z tests apply to test research hypotheses.

12.2.3 Multiple Regression

Research hypotheses that focus on a determination of the relationship between relevant health care measures and potential predispositional factors fall within the framework of multiple regression analysis. The general multiple regresssion model takes the form:

$$Y = XB + E \tag{12-15}$$

where

Y is an n by 1 vector representing sample observations of the dependent variable,

X is an n by p matrix of sample data for the (p-1) predispositional variables and intercept term,

B is a p by 1 vector of parameters to be estimated, and

E is an n by 1 vector of error terms.

The classical assumptions associated with the model are:

- The X matrix is composed of nonstochastic terms. In addition, no exact linear relationship exists among two or more of the independent variables.
- The dependent variable is normally distributed with mean XB and constant variance $I\sigma^2$, where I is an (n by n) identity matrix.
- The vector E consists of independent, identically distributed error terms which follow a normal distribution with zero mean and constant variance $I\sigma^2$.

Since the sampling strategy considered in the NMCES employed a differential probability of selection scheme, parameter estimates for the regression model must be derived using weighted least squares. Consequently, the matrix of estimated model parameters, B, take the form

$$B = (X' \; W \; X)^{-1} \; X' \; W \; Y \tag{12-16}$$

where W is an nxn diagonal matrix of analysis weights associated with the n sample observations.

The properties of the parameter estimates derived from probability weighted estimation are discussed in Holt, et al (1980), Kish and Frankel (1974), and Shah, et al (1977). Survey design complexities necessitate that variances of estimated model parameters be estimated by either jack-knife, balanced replication, or Taylor Series linearization methods. Hypothesis testing for model parameters can then be conducted by application of t-tests for individual parameters, or F-tests for multivariate considerations.

12.2.4 Analysis of Variance

Analysis of variance allows for the partitioning of the total variation in a criterion variable into several components. The analysis requires the specification of a set of factors or predispositional measures, each having several levels. The analysis determines the magnitude of the contributions of each of the factors to the total variation. It facilitates the estimation of domain means, defined by cross-classification of the factors, and hypothesis testing to determine which factors or interaction of factors make statistically significant contributions to the domain mean variation.

The classical analysis of variance framework assumes a normally distributed response variable, equality of variances, and zero covariances among domains, in addition to simple random sampling. F-tests are used to test research hypotheses. The properties of a complex survey design, however, significantly depart from the classical analysis of variance assumptions. To accommodate design complexities, the mean estimate for each domain is formulated as a ratio estimate that incorporates the analysis weights. The variance-covariance structure for the set of these domain means must be incorporated into the analyses of variance tests, when attempting to discern which sources of variation are statistically significant.

One approach to test relevant hypotheses is the weighted least squares methodology described by Grizzle, et al (1969) and implemented in the analysis of data from complex surveys by Koch, et al (1975), Freeman, et al (1976), Koch and Stokes (1980), Koch, et al (1980), and Landis, et al (1982). The strategy involves the specification of domains means in vector format, with a valid and consistent estimate of the related variance-covariance matrix, and the framework of a general linear model. The method is analogous to multiple regression analysis using weighted least squares for estimation of model parameters and for hypothesis testing. Because sample sizes are relatively large, it can be assumed that the estimated parameters have an approximate multivariate normal distribution. Consequently, statistical inferences can be made for model parameters by computing Wald statistics that have approximate chi-squared distributions. This allows for a determination of the factors or interaction of factors that make statistically significant contributions to the variation among domain means. The GENCAT program, a computer program for the generalized chi-square analysis of categorical data (Landis, et al, 1976), is particularly well suited for performing the relevant tests.

12.2.5 Contingency Table Analysis

Multi-dimensional contingency table analysis is appropriate when a criterion variable is defined in terms of categorical response profiles and further classified by related independent variables. The cross-classifications of the categorically defined independent variables define the mutually exclusive subpopulations. Inferences are made regarding the relationship between the independent factors and the response variable. Consider a response variable with r distinct response categories. Further, assume that a cross-classification of categorical independent variables define s distinct subpopulations that are of analytical interest.

When a simple random sampling scheme is considered, unbiased estimates of the subpopulation response profiles can be directly estimated from the respective frequency distributions characterizing the sample. The contingency table distribution is illustrated in the following table where

$n = \sum_i n(i\cdot)$ is the total sample size,

$n(i\cdot)$ represents the sample size for subpopulation i, and

$n(ij)$ represents the number of sample respondents in the i-th subpopulation with the j-th response (Grizzle et al, 1969).

Consequently, $P(ij)$ is an unbiased estimate of the proportion of the population in subpopulation i with response j where

$$P(ij) = n(ij) / n(i\cdot) \qquad (12\text{-}17)$$

Hypotheses are tested regarding the relationship between the factors and the response measure, using appropriate categorical data analysis techniques (Grizzle et al, 1969; Landis et al, 1982).

Subpopulation	Distinct Classes of Response						Sample Size
	1	2	· · ·	j	· · ·	r	
1	n(11)	n(12)	· · ·	n(ij)	· · ·	n(1r)	n(1·)
2	n(21)	n(22)	· · ·	n(2j)	· · ·	n(2r)	n(2·)
·	·	·	· · ·	·	· · ·	·	·
·	·	·	· · ·	·	· · ·	·	·
·	·	·	· · ·	·	· · ·	·	·
i	n(i1)	n(i2)	· · ·	n(ij)	· · ·	n(ir)	n(i·)
·	·	·	· · ·	·	· · ·	·	·
·	·	·	· · ·	·	· · ·	·	·
·	·	·	· · ·	·	· · ·	·	·
s	n(s1)	n(s2)	· · ·	n(sj)	· · ·	n(sr)	n(s·)

For complex survey designs similar to the NMCES, it is necessary to incorporate the analyses weights in the estimation strategy, to derive approximately unbiased national estimates for the response profiles within specified subpopulations. Application of the analysis weights in the estimation of response profiles is reflected in the alternative multi-dimensional contingency table where W(ijk) is the analysis weight for the k-th individual with the j-th response from the i-th subpopulation, $W(ij) = \sum_k W(ijk)$, the national estimate for subpopulation i with response j, and $W(i\cdot) = \sum_j W(ij)$ is the national population estimate for subpopulation i.

Subpopulation	Distinct Classes of Response						Population Estimate
	1	2	...	j	...	r	
1	W(11)	W(12)	...	W(1j)	...	W(1r)	W(1·)
2	W(21)	W(22)	...	W(2j)	...	W(2r)	W(2·)
.	.	.	...	.	...	.	.
.	.	.	...	.	...	.	.
.	.	.	...	.	...	.	.
i	W(i1)	W(i2)	...	W(ij)	...	W(ir)	W(i·)
.	.	.	...	.	...	.	.
.	.	.	...	.	...	.	.
.	.	.	...	.	...	.	.
s	W(s1)	W(s2)	...	W(sj)	...	W(sr)	W(s·)

Consequently, national estimates of the proportion of the population in subpopulation i with response j is determined as

$$P(ij) = W(ij) / W(i\cdot) \tag{12-18}$$

Estimates of the variances and covariances of the P(ij) terms for hypothesis testing are derived by a Taylor Series linearization approximation or application of replication strategies. The weighted least squares procedure presented in the section on analysis of variance can then be applied to the (sr by 1) vector of estimated proportions (expressed as ratio means) and associated variance-covariance matrix, to test research hypotheses.

12.2.6 Logistic Regression

Special consideration is required for the analysis of complex survey data, when the dependent variable in a multivariate analysis is binary. A dichotomous response variable in traditional regression analysis violates the assumptions of normality and continuity of the dependent variable. An appropriate analysis strategy for this type of data views the outcomes of the dependent variable as a probabilistic event, and the coefficients of the relevant model as the marginal changes in probability associated with each of the independent variables. The application of logistic regression analysis requires that the dependent variable be expressed in terms of the probability of an outcome [P(i)] for values of the explanatory variables associated with the i-th observation. The model takes the form:

$$\mathrm{Ln}\{P(i)/[1-P(i)]\} = B_0 + B_1X_1 + B_2X_2 + \ldots + B_nX_n + E_i \tag{12-19}$$

Statistical programs exist for logistic regression analysis of dichotomous data under simple random sampling assumptions, whether the data are grouped by categorical definitions for the independent variables or in original form. The GENCAT procedure, which uses a weighted least squares strategy to estimate model coefficients and test hypotheses of interest, is appropriate for logistic regression analysis when the independent variables are categorically defined (Landis, et al, 1976). Similarly, the method of maximum likelihood is used to estimate model coefficients for ungrouped data in the logistic regression program LOGIT (Nerlove and Press, 1973). However, the analytical requirements of a complex survey design are not directly satisfied by these programs. For grouped data, the GENCAT procedure can be modified for use with complex survey data by specifying the appropriate values for the variance-covariance matrix associated with the estimated log-odds ratios. When the ungrouped data are considered, however, the maximum likelihood LOGIT procedure has not been extended to derive the appropriate variance for model coefficients, or to test research hypotheses (Cohen, 1983).

12.3 Summary

Problems of variance estimation from complex survey designs have been discussed with particular reference to the NMCES survey. Several alternative variance estimators were reviewed, in particular, balanced repeated replication, the jack-knife method, and the Taylor Series linearization method. These variance estimation procedures incorporate the components of a complex survey design in their formulation. Attention was also given to the identification of existing statistical software, which produce appropriate variance estimates for means, totals, ratio estimates and linear regression coefficients.

Analytical strategies that incorporate survey design complexities are also presented. These include tests of hypotheses for population parameters, multivariate analyses within a regression or analysis of variance framework, and contingency table analyses. Available statistical package programs for their implementation are also referenced. Application of the analysis strategies that are appropriate for use with complex survey data serves to strengthen the validity of inferential test results.

References

Bean, J. A. (1975). Distribution and Properties of Variance Estimators for Complex Multistage Probability Samples. Vital and Health Statistics, DHEW Publication No. (HRA) 75-1339, Series 2, No. 65, Washington, DC: U.S. Government Printing Office.

Brillinger, D. R. (1964). The Asymptotic Behavior of Tukey's General Method of Setting Approximate Confidence Limits (the Jackknife) When Applied to Maximum Likelihood Estimates. Review of the International Statistical Institute, 3, 202-206.

Brillinger, D. R. (1966). The Application of the Jack-Knife to the Analysis of Sample Surveys. Commentary, 8, 74-80.

Brillinger, D. R. and J. W. Tukey (1964). Asymptotic Variances, Moments, Cumulants, and Other Average Values. Memorandum, Princeton, NJ: Princeton University.

Cohen, S. B. (1983). Present Limitations in the Availability of Statistical Software for the Analysis of Complex Survey Data. Review of Public Data Use, 11, 338-344.

Deming, W. E. (1956). On Simplification of Sample Design Through Replication With Equal Probabilities and Without Stages. Journal of the American Statistical Association, 51, 24-53.

Folsom, R. E. Jr. (1974). National Assessment Approach to Sampling Error Estimation, Sampling Error Monograph 25U-796-5 prepared for the National Assessment of Educational Progress (revised 1977).

Francis, I. (1983). A Survey of Statistical Software. *Computational Statistics and Data Analysis*, 1, 17-27.

Francis, I. and K. N. Berk (1978). A Review of the Manuals for BMD and SPSS. *Journal of the American Statistical Association*, 73, 65-71.

Frankel, M. R. (1971). *Inference from Survey Samples*, Ann Arbor, MI: Institute for Social Research, University of Michigan.

Freeman, D. H., J. L. Freeman, D. B. Brock, and G. G. Koch (1976). Strategies in the Multivariate Analysis of Data From Complex Surveys. *International Statistical Review*, 44, 317-330.

Fuller, W. A. (1975). Regression Analysis for Sample Surveys. *Sankhya C*, 37, 117-132.

Gurney, M. (1969). Random Group Method for Estimating Variances. Unpublished manuscript, U.S. Bureau of the Census, Washington, DC.

Grizzle, J. E., C. F. Starmer, and G. G. Koch (1969). Analysis of Categorical Data by Linear Models. *Biometrics*, 25, 489-504.

Helwig, J. T. and K. A. Council (1979). *SAS User's Guide*, 1979 Edition, Raleigh, NC: SAS Institute, Inc.

Holt, D., T. M. F. Smith, and P. D. Winter (1980). Regression Analysis of Data from Complex Surveys. *Journal of the Royal Statistical Society*, 143(4), 474-487.

Holt, M. M. (1977). *SURREGR: Standard Errors of Regression Coefficients From Sample Survey Data*, Research Triangle Institute, Research Triangle Park, North Carolina.

Jones, G. (1977). *Health Examination Survey (HES) Variance and Cross-tabulation Program - Version 2*, Internal Report, National Center for Health Statistics, Hyattsville, Maryland.

Jones, H. L. (1974). Jack-knife Estimation of Functions of Strata Means. *Biometrika*, 61, 343-348.

Keyfitz, N. (1957). Estimates of Sampling Variance Where Two Units Are Selected from Each Stratum. *Journal of the American Statistical Association*, 52, 503-510.

Kish, L. and M. R. Frankel (1974). Inferences from Complex Surveys. *Journal of the Royal Statistical Society*, 36, 1-37.

Koch, G. G., D. H. Freeman Jr., and J. L. Freeman (1975). Strategies in the Multivariate Analysis of Data From Complex Surveys. *International Statistical Review*, 43, 59-78.

Koch, G. G. and M. E. Stokes (1980). Annotated Computer Applications of Weighted Least Squares Methods for Illustrative Analysis of Examples Involving Health Survey Data. Technical Report prepared for the National Center for Health Statistics, Hyattsville, Maryland.

Koch, G. G., M. E. Stokes, and D. Brock (1980). Applications of Weighted Least Squares Methods for Fitting Variational Models to Health Survey Data. Proceedings of the American Statistical Association, Survey Research Methods Section, 218-223.

Landis, J. R., J. M. Lepkowski, S. A. Eklund, and S. A. Stehouwer (1982). A Statistical Methodology for Analyzing Data from a Complex Survey: The First National Health and Nutrition Examination Survey. Vital and Health Statistics Series, Series 2, No. 92, DHHS Publication No. (PHS) 821366, Washington, DC: U.S. Government Printing Office.

Landis, J. R., W. M. Stanish, J. L. Freeman, and G. G. Koch (1976). A Computer Program for the Generalized Chi-square Analysis of Categorical Data Using Weighted Least Squares (GENCAT). Computer Programs in Biomedicine, 6, 196-231.

Mahalanobis, P. C. (1939). A Sample Survey of the Acreage Under Jute in Bengal. Sankhya, 4, 511-531.

McCarthy, P. J. (1966). National Center for Health Statistics: Replication: An Approach to the Analysis of Data From Complex Surveys. Vital and Health Statistics, Series 2, No. 14, PHS Publication No. 79-1269, Washington, DC: U.S. Government Printing Office.

Nerlove, M. and S. J. Press (1973). Univariate and Multivariate Log-Linear and Logistic Models, Rand Report No. R-1306-EDA/NIH, Rand Corporation, Santa Monica, California.

Plackett, R. L. and J. P. Burman (1946). The Design of Optimum Multi-Factorial Experiments. Biometrika, 33, 305-325.

Shah, B. V. (1979). SESUDAAN: Standard Errors Program for Computing of Standardized Rates from Sample Survey Data, Research Triangle Institute, Research Triangle Park, North Carolina.

Shah, B. V. (1981). RATIOEST: Standard Errors Program for Computing Ratio Estimates for Sample Survey Data, Research Triangle Institute, Research Triangle Park, North Carolina.

Shah, B. V., M. M. Holt, and R. E. Folsom (1977). Inferences About Regression Models From Sample Survey Data, Bulletin of the International Statistical Institute, XLVII(3), 43-57.

Tepping, B. J. (1968). The Estimation of Variance in Complex Surveys. *Proceedings of the American Statistical Association, Social Statistics Section*, 11-18.

Tukey, J. W. (1958). Bias and Confidence in Not Quite Large Samples. *Annals of Mathematical Statistics*, 29, 614.

Van Eck, N. (1979). *OSIRIS IV User's Manual*, Fifth Edition, Ann Arbor: Institute for Social Research, The University of Michigan.

Woodruff, R. S. (1971). A Simple Method for Approximating the Variance of a Complicated Estimate. *Journal of the American Statistical Assciation*, 66, 411-414.

13
Variance Approximation Strategies for Survey Estimates

Data from complex survey designs require special procedures for variance estimation and analysis, as a consequence of design components that include unequal selection probabilities, stratification, and area sampling. With this deviation from simple random sampling, many of the basic statistical packages (e.g., SAS, SPSS) can no longer provide appropriate variance estimates for survey statistics. Specially designed software packages do exist for direct variance estimation of complex survey statistics such as means, totals, ratios, proportions, and regression coefficients (SESUDAAN, SURREGR, OSIRIS). The variance estimation methods used by these statistical packages include balanced repeated replication, jack-knife, and Taylor Series linearization. These procedures are described in the previous chapter.

When many tables are required, the computation cost can be large if these procedures are used for each estimate of interest. For this reason, variance estimates cannot always be computed for each statistic by direct methods. In addition, the inclusion of variance estimates for all statistics in a data report would yield cumbersome documents in some instances. Consequently, generalized variances are often considered as an alternative approach. Generalized variances are model-based estimates of the variances of survey statistics, which require that variance estimates be derived by direct methods for only a subset of all parameter estimates being computed. The subset should be characterized by a range of statistics and domains to insure a representative mix of variability in the parameter estimates.

This chapter focuses on two generalized variance estimation approaches that incorporate the complexities of the survey design in a cost efficient manner: the design effect model and the relative variance curve strategy. The model specifications for these two alternative variance estimation techniques are presented in detail. In addition, the accuracy of the alternative variance estimates are compared for survey statistics from the National Medical Care Expenditure Survey. The survey statistics included in the study were estimates of the medical care utilization, expenditures, and health insurance coverage characteristics of the population of the United States. Two design effect procedures were compared to the relative variance curve procedure for accuracy of the predicted variances: the average design effect model stratified by sample size and criterion variable boundaries and the average design effect model without stratification (Cohen, 1982, 1983).

13.1 The Average Design Effect Model

The design effect method is a commonly used alternative to direct variance estimation for survey statistics. The design effect (DEFF) is defined as the ratio of the true variance of a statistic $\hat{\mu}$ to the variance derived under simple random sampling (SRS) assumptions, or

$$\text{DEFF}(\hat{\mu}) = \text{Var}(\hat{\mu}) / \text{SRS Var}(\hat{\mu}) \tag{13-1}$$

where $\text{Var}(\hat{\mu})$ is the variance of $\hat{\mu}$ and SRS $\text{Var}(\hat{\mu})$ is the variance that would have resulted from simple random sampling. The average design effect model for generalized variance estimation computes the average of the design effect for a subset of survey statistics. For statistics for which a generalized variance is desired, variances are derived under simple random sampling assumptions and adjusted by multiplying by the average design effect to produce the generalized variance estimate.

For those data bases originating from complex national survey designs, it is not unusual to have sample sizes in excess of 10,000 individuals. A considerable reduction in computation costs may be achieved for these large data bases when variances are estimated under SRS assumptions using statistical package programs, as an alternative to the Taylor Series linearization method, balanced repeated replication or the jack-knife method.

Since the design effect is generally a stable measure, with a much narrower range of dispersion than direct variance estimates, use of an average design effect to compute generalized variance estimates offers a reasonably good trade-off in cost savings for the attendant reduction in accuracy. The accuracy of the average design effect method may be inferred from the level of dispersion characterizing the design effects for a set of related statistics.

The introduction of stratification procedures to the average design effect methodology is one potential strategy for achieving gains in precision for the resultant estimates. The strategy has been suggested by Kish and Frankel (1974), as an alternative to the process of directly computing sampling errors for all the different survey variables under investigation and for their respective subclasses. Direct estimates of the design effects are computed for a subset of statistics, and their mean applied to the entire set of related statistics. Since design effects vary as the statistics, variables, and survey designs differ, the strategy should be separately applied for groups of related variables for the particular survey at hand.

In applying the technique, a criterion variable of interest is identified (e.g., medical expenditures) and several domain estimates of this criterion variable are produced. The domain estimates are defined by marginal and cross-classified distributional categories of predetermined demographic measures (e.g., mean annual medical expenditures for specific age-race-sex-income classes of the national population). For a representative subset of the domain estimates to be made for the criterion variable, direct estimates of the design effects are derived. An average design effect (ADEFF) is then determined in the following manner:

$$ADEFF = \sum_{d=1}^{D} DEFF(d)/D \qquad (13\text{-}2)$$

where

DEFF(d) is a direct estimate of the design effect for domain estimate d,

D is the number of domain estimates selected in the representative subset, and

ADEFF is the average design effect.

A weighted average design effect (WDEFF) may also be used, where each design effect is weighted by the population estimate for the domain it represents. Here,

$$WDEFF = \sum_{d=1}^{D} \hat{Y}(d)\ DEFF(d) \Big/ \sum_{d=1}^{D} \hat{Y}(d) \qquad (13\text{-}3)$$

where $\hat{Y}(d)$ is the estimated population total for domain D.

Once the average design effect is determined, variance estimates for all related statistics derived under SRS assumptions can be adjusted by this factor, to produce the appropriate variance estimate. The cost savings in computer time and dollars may be substantial, considering the data permutations that may be relevant for a diverse population.

13.2 Relative Variance Curve Strategy

The _relative variance curve_ approach to generalized variance estimation is to develop a functional model for the relationship between the relative variance of a statistic and the statistic itself. The _relative variance_ of an estimator $\hat{\mu}$ for a population parameter μ, is defined as the variance of the statistic divided by the square of the parameter being estimated. Using the estimator $\hat{\mu}$ and its estimated variance Var($\hat{\mu}$), the following estimate of the relative variance (Rel Var) may be obtained:

$$\text{Rel Var}\ (\hat{\mu}) = \text{Var}\ (\hat{\mu})\ /\ \hat{\mu}^2 \tag{13-4}$$

The relative variance curve approach for a particular type of survey estimate can best described in a stepwise manner. First, estimates and their relative variance are computed using direct estimation procedures that account for the complex survey design. Then a functional model is developed for the relative variance in terms of the parameter estimate or

$$\text{Rel Var}\ (\hat{\mu}) = f(\hat{\mu}) \tag{13-5}$$

The model is then tested and revised to produce the best fitting model for predicting the relative variance of an estimate given the estimate. When the variance is needed for a statistic of the same type, say $\hat{\mu}'$, the relative variance [rel var($\hat{\mu}'$)] is approximated using this model. Then the variance is estimated as

$$\text{var}\ (\hat{\mu}') = [\text{rel var}\ (\hat{\mu}')]\ (\hat{\mu}')^2 \tag{13-6}$$

A standard error estimate is obtained as the square root of the variance estimate or

$$\text{se}\ (\hat{\mu}') = [\text{rel se}\ (\hat{\mu}')]\ (\hat{\mu}') \tag{13-7}$$

where the relative standard error of $\hat{\mu}'$ or rel se ($\hat{\mu}'$) is computed as square root of the relative variance or

$$\text{rel se } (\hat{\mu}') = [\text{rel var } (\hat{\mu}')]^{\frac{1}{2}} \qquad (13\text{-}8)$$

For reporting, it is often more convenient to provide relative standard error curves, since equation (13-7) can then be directly used to compute an approximation for the standard error of the estimate.

In addition to savings obtained in variance computation costs, the inclusion of summary relative standard error tables in statistical reports results in savings in production costs as an alternative to publication of a variance estimate for each statistic that is presented. To illustrate the presentation of these summary tables, Table 13-1 contains approximate relative standard errors (expressed

Table 13-1. Example of a Relative Standard Error Table for Estimated Population Totals

Estimated Population Totals (in thousands)	Relative Standard Error (%)
500	18.9
1,000	13.5
2,500	8.7
5,000	6.4
10,000	4.8
25,000	3.5
50,000	3.0
100,000	2.6
200,000	2.4

as a percent) for estimated population totals. As an example of the use of Table 13-1, an estimate of 26.3 million persons in the U.S. with at least one purchase or repair of eyeglasses or contact lenses has a relative standard error of approximately 3.5 percent. The standard error of this estimate is obtained as:

$$se(\hat{Y}) = 26{,}300{,}000\ (3.5)/100 = 920{,}500$$

The model that is used to predict the relative variance will depend upon whether a population total, a proportion, or a mean is being estimated. These models are presented in the remainder of this section, together with methods for fitting the models.

13.2.1 Variance Curves for Aggregate Statistics

Often estimates are needed of population counts (e.g., number of persons living below the poverty line, number of persons burdened with high medical expenses) or other population aggregates (e.g., total expenditures for medical care, total number of physician visits made in a particular year). Let $\hat{Y}$ be an estimated population aggregate obtained from a sample survey or

$$\hat{Y} = \sum_{i \varepsilon S} W(i)\ Y(i) \tag{13-9}$$

where W(i) is the analysis weight associated with the i-th member of the sample, Y(i) is the response of the i-th sample member, and $i \varepsilon S$ denotes that the summation is over all sample members belonging to the sample (S). If a population total is desired for domain d, then Y(i) will be "1" when sample member i belongs to domain d and "0" otherwise. When total utilization or expenditures for the nation are required, Y(i) will be the utilization or expenditures made by the i-th sample member.

For aggregate statistics, the relative variance curve strategy assumes that there is an inverse relationship between the size of an estimate $\hat{Y}$ and its relative variance (Bean, 1970). The assumed relationship between Rel Var$(\hat{Y})$ and $\hat{Y}$ is expressed as:

$$\text{Rel Var } (\hat{Y}) = \alpha + \beta/\hat{Y} + \varepsilon \qquad (13\text{-}10)$$

where α is the intercept and β the slope of the regression model and ε is a random error term associated with a particular estimate. Based upon this model, the relative variance is approximated as

$$\text{rel var } (\hat{Y}) \doteq a + b/\hat{Y} \qquad (13\text{-}11)$$

where a and b are estimates for α and β respectively. These estimates can be determined by an iterative or weighted least squares procedure (Cohen, 1979).

To illustrate the method, a relative variance model was developed for estimates of the number of insured persons from different demographic subgroups of the national population. The relative standard error curve is shown in Figure 13-1 where the estimated values of α and β are a = 0.001732 and b = 32,311.84. To explain the use of this curve, suppose the standard error of an estimate of ten million insured persons is needed. First, ten million would be located on the bottom of the chart and the intersection of the relative standard error curve found. The intersection occurs at

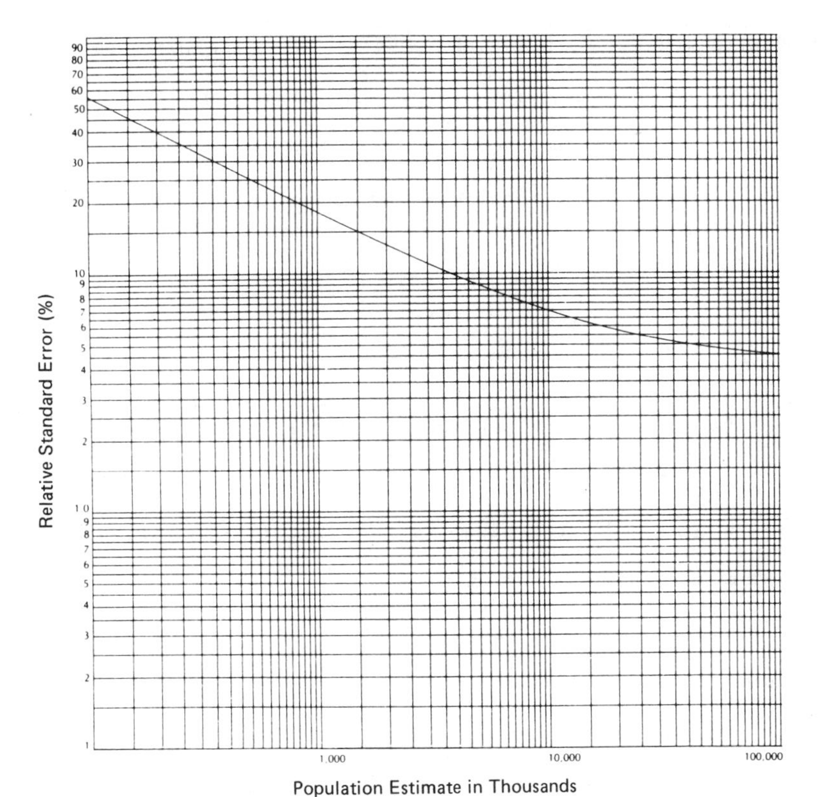

Figure 13-1. Smoothed Relative Standard Error Curve for Aggregate Estimates of the Insured Population (Round 1 of NMCES). (From Steven B. Cohen and William D. Kalsbeek, National Health Care Expenditures Study NCHSR 81-149, U.S. Department of Health and Human Services, Hyattsville, Maryland, June 1981.)

seven percent indicating a relative standard error of seven percent of ten million for a standard error of 700,000 (seven percent of ten million).

13.2.2 Variance Curves for Population Percentages

Another class of statistics for which variance estimates are frequently needed are population percentages (e.g., the percent of the population that live below the poverty line, the percentage of total medical expenses associated with hospital stays). For statistics of this type, the numerator is a subclass of the denominator. The sample estimate for a population percentage P would be obtained as

$$\hat{P} = 100 \left[\sum_{i \varepsilon S} \delta_d(i)\, W(i)\, Y(i) \right] / \left[\sum_{i \varepsilon S} W(i)\, Y(i) \right] \qquad (13\text{-}12)$$

where W(i) is again the analysis weight of the i-th sample member, Y(i) is the response of the i-th sample member, and $\delta_d(i)$ is a domain membership indicator equal to "1" when sample member i belongs to domain d and "0" otherwise.

An illustration of the equation (13-12) is in order. Suppose the statistic of interest is the percentage of the population living below the poverty line. Then $\delta_d(i)$ would be "1" if the i-th sample person lived below the poverty line and "0" otherwise; the summations would be over all sample persons and Y(i) would be "1" for all persons. As another example, the estimate of the percentage of total medical expenses associated with hospital stays would be constructed by summing over all medical events associated with sample persons. The indicator $\delta_d(i)$ would be "1" if the event were a hospital stay and "0" otherwise. The variable Y(i) would be the expense associated with the i-th event.

Note that equation 13-12 can be written more simply as

$$\hat{P} = 100\, \hat{Y}(d) / \hat{Y} \qquad (13\text{-}13)$$

where $\hat{Y}(d)$ is the estimated total for domain d members and $\hat{Y}$ is the estimated total for the entire population. The relative variance curve approach to variance estimation assumes that the relative variances for the two population estimates, $\hat{Y}(d)$ and $\hat{Y}$, can be derived from the same model for a population total. When the correlation between $\hat{P}$ and $\hat{Y}$ is negligible, the Taylor Series approximation for the relative variance of $\hat{P}$ will be

$$\text{Rel Var}(\hat{P}) \doteq \text{Rel Var}[\hat{Y}(d)] - \text{Rel Var}(\hat{Y}) \tag{13-14a}$$

If equation (13-10) holds for both $\hat{Y}(d)$ and $\hat{Y}$, then the relative variance of $\hat{P}$ can be approximated as

$$\text{rel var}(\hat{P}) \doteq b/\hat{Y}(d) - b/\hat{Y} \tag{13-14b}$$

Noting that $\hat{Y}(d) = \hat{P}/100$, the following more convenient estimate may be obtained for the relative variance:

$$\text{rel var}(\hat{P}) \doteq (b/\hat{Y})\ [(100-\hat{P})/\hat{P}] \tag{13-15}$$

where b is the estimated coefficient determined in the curve fitting procedure for aggregate statistics. Consequently, the relative variance of percentage estimates depends upon both the respective population base Y and the percent value P.

To illustrate the procedure, Figure 13-2 presents the relative standard error curves for percentages based upon characteristics of the insured population. Linear interpolation is needed when the population base of the percent is between values specified for the relative standard error curves. As an example of the use of the

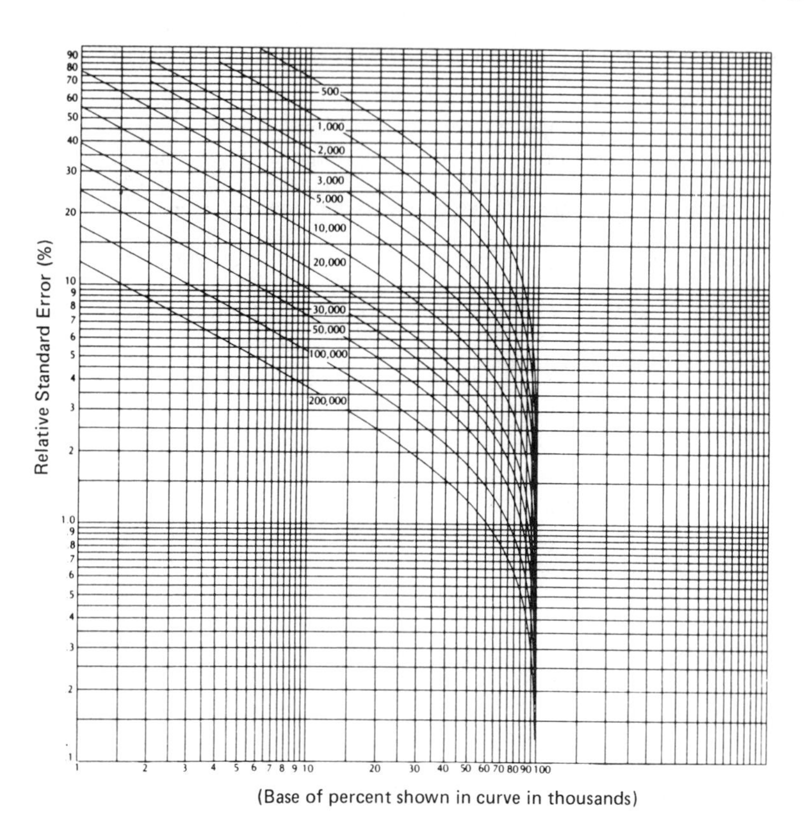

Figure 13-2. Relative Standard Error Curves for Percent of Insured Population (Round 1 of NMCES). (From Steven B. Cohen and William D. Kalsbeek, National Health Care Expenditures Study NCHSR 81-149, U.S. Department of Health and Human Services, Hyattsville, Maryland, June 1981.)

chart, an estimate of ten percent (the scale at the bottom of the chart) from a population of 10 million insured persons (sixth curve from top) would have an estimated relative standard error of 16.8 percent or a standard error of 1.68 percent (16.8 percent of 10 percent).

13.2.3 Variance Curves for Ratio Estimators

The final type of statistics that are commonly estimated using survey data is ratio estimates such as means. Ratio estimates differ from percentages in that the numerator is not a subgroup of the denominator. Examples of ratio estimates would be the average employment income of workers in the United States or the average charge per hospital stay. The parameter of interest is the ratio of two population totals or $R = X / Y$. For the average employment income of residents of the United States, X would be the total employment income received by all workers in the United States, and Y would be the total number of workers. For the average charge per hospital stay, X would be the total charges associated with all hospital stays and Y the total number of hospital stays.

Using survey data, the estimator $\hat{R}$ for a ratio estimate is obtained as:

$$\hat{R} = \hat{X} / \hat{Y} \tag{13-16}$$

where $\hat{X}$ is the sample estimate of population total X and $\hat{Y}$ is the sample estimate of population total Y. The estimates for $\hat{X}$ and $\hat{Y}$ may be obtained using equation (13-9).

Generalized variances of ratio estimators are based upon the assumption that the covariance between $\hat{X}$ and $\hat{Y}$ is negligible so that the relative variance of the ratio is approximately equivalent to the sum of the relative variances of the numerator and denominator (French, 1978). This relation takes the form:

$$\text{Rel Var}(\hat{R}) \doteq \text{Rel Var}(\hat{X}) + \text{Rel Var}(\hat{Y}) \tag{13-17}$$

The relative variance curve can be approximated as:

$$\text{rel var}(\hat{R}) = a_1 + b_1/\hat{X} + a_2 + b_2/\hat{Y} \tag{13-18}$$

A more convenient form of the relative variance is obtained by noting that $\hat{X} = \hat{R}\ \hat{Y}$ or

$$\text{rel var}(\hat{R}) = a + b_1\ /(\hat{R}\ \hat{Y}) + b_2/\hat{Y} \tag{13-19}$$

where $a = a_1 + a_2$.

Note that the relative variance of the ratio estimator is inversely related to the size of the population base Y and the ratio value R.

13.2.4 Methods for Estimating Model Coefficients

Several alternative curve fitting procedures with different optimization criteria have been suggested for estimating model coefficients. These include an iterative procedure that minimizes the relative squared derivations of predicted versus observed relative variances and a weighted least squares estimation strategy.

The iterative procedure described by Bean (1970) produces estimates of a and b that minimize the squared relative residuals of the model-based relative variance [rel var($\hat{\mu}$)] from the survey estimated value [Rel Var($\hat{\mu}$)]. This process will be illustrated for the development of a model for a population total.

Suppose that for I estimated totals the relative variance (Rel Var) has been calculated by direct methods. Then the squared relative residual between the directly estimated values and the model predicted values would be expressed as:

$$SSR = \{ \sum_{i=1}^{I} \{Rel\ Var[\hat{Y}(i)] - [a + b / \hat{Y}(i)]\} / True\ Rel\ Var[\hat{Y}(i)]\}^2 \tag{13-20}$$

where Rel Var$[\hat{Y}(i)]$ is the relative variance for the i-th statistic ($i = 1,2,\ldots,I$) estimated from survey data, and True Rel Var$[\hat{Y}(i)]$ the unknown true relative variance. Starting values of a and b are derived by considering the normal equations for SSR, where the True Rel Var$[\hat{Y}(i)]$ is approximated by Rel Var$[\hat{Y}(i)]$. Once values for a and b are determined, $a + b / \hat{Y}(i)$ is computed and substituted for True Rel Var$[\hat{Y}(i)]$. This allows for the computation of new estimates of a and b. This procedure continues until the estimates from the j-th and (j-1)-th iteration converge or

$$|(a_j - a_{j-1})/a_j| < 0.01$$

and

$$|(b_j - b_{j-1})/b_j| < 0.01$$

Because the empirical relationship between relative variances and aggregate estimates (totals) is linear in parameters, a least squares technique for estimating α and β in equation (13-10) is also appropriate. Consider the linear transformation $Z(i) = 1/\hat{Y}(i)$ so that

$$Rel\ Var[\hat{Y}(i)] = \alpha + \beta\ Z(i) + \varepsilon \tag{13-21}$$

When this model was separately applied to NMCES data on health care utilization, expenditures, and insurance coverage, residuals were examined. The findings suggest the variance of the Rel $\mathrm{Var}[\hat{Y}(i)]$ varies inversely as the size of the aggregate estimate $[\hat{Y}(i)]$ increases [or directly as Z(i) increases], so that

$$\mathrm{Var}\{\mathrm{Rel\ Var}[\hat{Y}(i)]\} = \sigma^2 Z(i) \tag{13-22}$$

thus violating the underlying assumption of homoscedasticity for ordinary least squares analysis (Cohen, 1979). Use of ordinary least squares in this setting would yield unbiased estimates of the regression coefficients but would not be efficient. Consequently, weighted least squares is preferred to produce the minimum variance unbiased estimates of these coefficients. The appropriate weights, W(i), are of the form

$$W(i) = Z(i) \tag{13-23}$$

The relationship in equation (13-10) may be re-expressed as:

$$\underset{\sim\sim}{RV} = \underset{\sim}{Z}\, \underset{\sim}{\beta} \tag{13-24}$$

where $\underset{\sim}{\beta} = (\alpha, \beta)'$, $\underset{\sim\sim}{RV}$ is the I by 1 vector of observed relative variances, and $\underset{\sim}{Z}$ is an I by 2 matrix with the first column all 1's and the second column the I values observed for Z(i). Then $\underset{\sim}{\beta} = (\alpha,\beta)'$ would be estimated by $\underset{\sim}{b} = (a,b)'$, where

$$\underset{\sim}{b} = (\underset{\sim}{Z}'\, \underset{\sim}{W}^{-1}\, \underset{\sim}{Z})^{-1}\, \underset{\sim}{Z}'\, \underset{\sim}{W}^{-1}\, \underset{\sim\sim}{RV} \tag{13-25}$$

The matrix $\underset{\sim}{W}$ would have the W(i) weights as its diagonal elements and zero for the off diagonal elements. Once the model coefficients are determined, variances can be predicted for all related statistics through application of a conversion factor to the resultant relative variance estimates. This conversion factor is the estimated statistic squared.

13.3 Comparison of the Accuracy Between Methods

To investigate the accuracy of these two alternative methods of variance estimation for the NMCES, comparisons were made for a representative set of survey statistics of the medical care utilization, expenditures, and health insurance coverage of the national population. The utilization measures included the number of physician visits, hospital admissions and number of prescribed medicines.* Expenditure statistics for each of these utilization measures were also included: physician visit expenditures, total expenditures for prescribed medicines, and total expenditures for all hospital admissions (with charges excluded for inpatient physician services). The measure of health insurance coverage indicated the presence of private health insurance coverage. In addition, the domain-defining demographic measures for the survey statistics under consideration included age (0 to 4, 5 to 14, 15 to 24, 25 to 34, 35 to 44, 45 to 54, 55 to 64, 65 or more), race (white, nonwhite), sex

---

*Physician visits consisted of all ambulatory physician contacts, excluding telephone calls. Hospital admissions included admissions of less than 24 hours and those for women giving birth. Newborns were not counted as separate admissions unless they were admitted separately following delivery. Prescribed medicines included any drug or other medical preparation prescribed by a physician, including refills.

(male, female), self-reported health status (excellent, good, fair, poor), marital status (never married, married, widowed, separated, divorced, less than 17 years of age), years of education (0 to 8, 9 to 11, 12, 13 to 15, 16 or more, less than 17 years of age), employment status (employed, unemployed, not in labor force, less than 16 years old) and size of city (SMSA, nonSMSA).

The diverse set of criterion variables also served to represent three distinct classes of survey statistics: narrow, medium and wide range. More specifically, "narrow range" statistics are usually associated with data at the individual level, and are based upon measurements generally falling within the range of 0 to 3. These measurements indicate the presence or absence of a population attribute or its frequency of occurrence. Narrow range statistics were represented by NMCES data on insurance coverage, and number of hospital admissions. "Medium range" statistics consist of measurements which infrequently fall outside the range of 0 to 10. Data on ambulatory visits and number of prescribed medicines served to represent the medium range class. "Wide range" statistics are characterized by data more continuous in nature, which have much higher upper bounds. Wide range statistics were represented by the following measures: total expenditures for hospital admissions, physician expenditures, and total expenditures for prescribed medicines.

For each of the selected criterion variables, domain estimates were generated in terms of population means or proportions when appropriate. The domain estimates were defined by marginal and cross-classified distributional categories of the selected demographic measures. For example, consider the mean annual expenditures for ambulatory physician visits within specific age, race, sex, health-status classes of the U.S. population. The domain d estimate, $\bar{y}(d)$, is derived as:

$$\bar{y}(d) = \sum_{i \varepsilon S} \delta_d(i)\ W(i)\ Y(i) \ / \ \sum_{i \varepsilon S} \delta_d(i)\ W(i) \qquad (13\text{-}26)$$

where

$Y(i)$ is the i-th individual's expenditures for physician visits,

$W(i)$ is the i-th individual's analysis weight, and

$\delta_d(i)$ is an indicator of domain d membership where $\delta_d(i) = 1$ if the i-th individual is a member of the d-th domain and 0 otherwise.

In this study, model parameters for the relative variance curve method were estimated by the weighted least squares optimization strategy. The unweighted estimator was used for the average design effect model. An examination of the design effects for domain estimates of the NMCES health care utilization, expenditure, and insurance coverage measures indicated that design effect variation was influenced by sample size and the size of the domain estimate (Cohen, 1983). Consequently, the stratification scheme was implemented by constructing quartile boundaries on sample size for the set of domain estimates under investigation, cross-classified by tertile boundaries on the size of the resultant domain estimates for the respective health care measure. Within each of these twelve distinct strata, the average design effect was computed and used for variance estimation. All direct variance estimates for survey statistics were generated through the Taylor Series linearization method. Modelling for each of the specified variance approximation strategies was done separately for the different sets of criterion variables.

To measure the accuracy of the variance estimation strategies, the average relative absolute difference was considered between direct and predicted estimates of variance for domain specific population estimates. The measure took the form:

$$\bar{A} = \sum_{d=1}^{D} \mid \hat{S}^2_p(d) - \hat{S}^2_0(d) \mid / \hat{S}^2_0(d) \qquad (13\text{-}27)$$

where $\hat{S}^2_0(d)$ is the observed direct variance estimate obtained by the Taylor Series linearization method for the d-th domain population estimate, $\hat{S}^2_p(d)$ is the predicted variance by either the average design effect or relative variance curve method for the d-th domain population estimate, and D is the number of domain estimates that constitute a representative subset for the criterion variable of interest.

Table 13-2 presents the comparisons of accuracy for the alternative variance estimation techniques: the relative variance curve strategy, the average design effect model, and the average design effect model with stratification. Study findings revealed a consistently lower average relative absolute difference for both design effect methods, over the relative variance curve technique. All differences in accuracy were significant at the 0.01 level as determined by application of paired t-tests. The null hypothesis of interest was specified as: H_0: no difference in accuracy versus H_a: a difference in accuracy. The improvements in accuracy were most notable for the prescribed medicine and physician-related parameter estimates. In addition, the comparison of the two average design effect models (with and without stratification) revealed significant improvements in accuracy were obtained through the introduction of stratification.

The lower order of magnitude in accuracy for the relative variance curve strategy was disturbing. The technique has gained a degree of respectability as a consequence of its theoretical justification and widespread useage among a large statistical audience. Given the convenience of providing a limited number of variance curves in study reports, many users would be willing to accept modest levels of bias when the variance curve strategy is applied. However, the consistently greater accuracy of the design effect estimation strategy argues that greater scrutiny must be given to the relative variance curve strategy prior to a decision for adoption.

Table 13-3 presents the relative percent reduction in the average absolute relative difference obtained by using the stratified average design effect model over the relative variance curve strategy or the unstratified average design effect model. This measure, I, which also indicates the relative percent improvement in accuracy of an improved estimation strategy S_{imp} over an alternative S_{alt} is expressed as:

Table 13-2. Comparison of Average Relative Absolute Difference for the Alternative Generalized Variance Strategies

Class of Statistics	Number of Domain Estimates	Average Relative Absolute Difference and Standard Error		
		Average Design Effect Model With Stratification	Average Design Effect Model	Relative Variance Curve Model
Narrow Range:				
Hospital Admissions	386	0.075 (0.003)	0.080 (0.003)	0.623 (0.028)
Private Insurance Coverage	386	0.197 (0.008)	0.354 (0.012)	0.556 (0.021)

Medium Range:				
Physician Visits	386	0.110 (0.004)	0.126 (0.005)	1.706 (0.059)
Prescribed Medicines	386	0.110 (0.004)	0.153 (0.006)	1.423 (0.053)
Wide Range:				
Hospital Expenditures	386	0.093 (0.004)	0.097 (0.004)	0.411 (0.023)
Physician Expenditures	386	0.093 (0.004)	0.112 (0.005)	1.221 (0.041)
Prescribed Medicines Expenditures	386	0.104 (0.004)	0.147 (0.005)	1.388 (0.043)

*All pairwise differences were significant at the 0.01 level, when testing the null hypothesis H_0: no difference in accuracy across models versus H_a: difference in accuracy.

Table 13-3. Percent Relative Reduction in Average Absolute Relative Difference and Ratios of Standard Errors for the Accuracy Measures*

Class of Statistics	Number of Domain Estimates	Percent Relative Reduction in $\bar{A}$			Ratio of Standard Errors		
		$\bar{A}$(DS) to $\bar{A}$(RV)	$\bar{A}$(DW) to $\bar{A}$(RV)	$\bar{A}$(DS) to $\bar{A}$(DW)	SE(DS) to SE(RV)	SE(DW) to SE(RV)	SE(DS) to SE(DW)
Narrow Range:							
Hospital Admissions	386	88.0	87.2	6.3	0.106	0.118	0.894
Private Insurance Coverage	386	64.6	35.3	44.4	0.390	0.580	0.672

Medium Range:							
Physician Visits	386	93.6	92.6	12.7	0.081	0.081	0.991
Prescribed Medicines	386	92.3	89.2	28.1	0.083	0.104	0.791
Wide Range:							
Hospital Expenditures	386	77.4	76.4	4.1	0.184	0.194	0.948
Physician Expenditures	386	92.4	90.8	17.0	0.101	0.116	0.877
Prescribed Medicine Expenditures	386	92.5	89.4	29.3	0.098	0.123	0.795

*A "DS" denotes the average design effect model with stratification, a "DW" denotes the average design effect model without stratification, and a "RV" denotes the relative variance curve model.

Table 13-4. Range in Values of Relative Absolute Difference in Variance Estimates Between Predicted and Directly Estimated Values

Class of Statistics	Number of Domains	Range in Value of Relative Absolute Difference: Design Effect Model With Stratification (lower, upper)	Design Effect Model Without Stratification (lower, upper)	Relative Variance Curve Model (lower, upper)
Narrow Range:				
Hospital Admissions	386	<0.001, 0.325	<0.001, 0.353	0.002, 2.767
Private Insurance Coverage	386	<0.001, 0.885	0.003, 1.166	<0.001, 2.562

Medium Range:				
Physician Visits	386	<0.001, 0.568	<0.001, 0.645	0.020, 5.896
Prescribed Medicines	386	<0.001, 0.425	<0.001, 0.520	<0.001, 4.448
Wide Range:				
Hospital Expenditures	386	<0.001, 0.446	<0.001, 0.453	0.001, 2.281
Physician Expenditures	386	<0.001, 0.661	<0.001, 0.756	0.006, 3.859
Prescribed Medicine Expenditures	386	<0.001, 0.424	<0.001, 0.524	0.038, 3.955

$$I = 100\ [\bar{A}(S_{alt}) - \bar{A}(S_{imp})]\ /\ \bar{A}(S_{alt}) \qquad (13\text{-}28)$$

where $\bar{A}(S_{alt})$ and $\bar{A}(S_{imp})$ are defined as the average relative absolute deviation in variance estimates for the respective estimation strategies.

The design effect models consistently yielded a reduction in $\bar{A}$ over the relative variance curve technique. For the average design effect model with stratification, the minimum relative reduction in $\bar{A}$ was 65 percent. For over 50 percent of the specified comparisons, the reduction was greater than 92 percent, signifying a marked improvement in accuracy. When comparisons were directed towards the percent improvement in accuracy obtained by the design effect model, the stratified model was consistently judged superior.

The ratio of standard errors for the accuracy measures of the respective variance estimation models are presented in Table 13-3. For each of the data sets under investigation, none of the observed ratios relative to the relative variance curve method exceeded 0.58. In addition, the ratio for the design effect model with stratification, relative to the overall average design effect model, was consistently less than one with stratification. Further, the investigation revealed the range of absolute relative differences between predicted and direct variance estimates were much narrower for the design effect model (Table 13-4). Consequently, this strategy also demonstrates superior performance in its capacity to yield variance estimates with the greatest precision. In this setting, precision was defined in terms of the range of relative absolute deviations between predicted and Taylor Series direct variance estimates for the diverse set of specified domains.

13.4 Summary

Several direct methods of variance estimation appropriate for complex survey data have been developed. These include the Taylor Series linearization method, balanced repeated replication, and the jack-knife method. Use of these procedures can become costly with respect to computer time, if applied to each parameter estimate of interest for a large number of tables. Consequently, some users are willing to use generalized variance estimation strategies if they result in only modest levels of bias.

This chapter described two alternative variance estimation strategies that have gained widespread use in the statistical community. The accuracy of these methods was compared for a representative set of survey statistics specific to data from the National Medical Care Expenditure Survey. The design effect model consistently yielded variance estimates that were superior in terms of accuracy and precision when compared with those derived by the relative variance curve strategy. Further gains in accuracy were achieved for the average design effect model with the introduction of stratification. The results demonstrate that the decision concerning the method for adoption should not be solely based upon convenience. Measures of accuracy should be defined, and the behavior of both methods compared for a representative subset of sample estimates. The method of variance estimation that displays the superior performance for the specified measures should then be used.

References

Bean, J. A. (1970). Estimation and Sampling Variance in the Health Interview Survey. Vital and Health Statistics, Series 2, No. 28, PHS Publication No. 1000, Washington, DC: U.S. Government Printing Office.

Cohen, S. B. (1979). An Assessment of Curve Smoothing Strategies Which Yield Variance Estimates for Complex Survey Data. Proceedings of the American Statistical Association, Survey Research Methods Section, 101-104.

Cohen, S. B. (1982). Comparison of Design Effect and Relative Variance Curve Strategy for Variance Estimation from Complex Survey Data. Presented in the 1982 annual meetings of the American Public Health Association and available from the National Center for Health Services Research, Rockville, Maryland.

Cohen, S. B. (1983). Design Effect Variation in the National Medical Care Expenditure Survey. Proceedings of the American Statistical Association, Survey Research Methods Section, 748-756.

French, D. K. (1978). National Survey of Family Growth, Cycle 1: Sample Design, Estimation Procedures, and Variance Estimation. Data Evaluation and Methods Research, Series 2, No. 76, PHS Publication No. 78-1350, Washington, DC: U.S. Government Printing Office.

Kish, L. and M. R. Frankel (1974). Inferences From Complex Surveys. Journal of the Royal Statistical Society, 36, 1-37.

14

Family Unit Analysis in the National Medical Care Expenditure Survey

Research to address social concerns frequently requires data at the family level as well as at the individual level. Interest in family-level data arises because the actions and attitudes of individuals are influenced by family characteristics. Family income prompts many decisions that individuals make, for instance. As another example, the burden associated with health care usage is a function of total family expenditures compared to total family resources. For many social surveys, family considerations must be incorporated into analyses.

Designing a cross-sectional or one time survey to obtain family and individual level data is usually straightforward. Only those individuals and families in existence on the date of the interview are included in the study. A definitional problem does arise with respect to the treatment of persons living temporarily away from home (e.g., college students, spouses away on work assignments). This latter complication is primarily associated with the definition of what constitutes a "family." Study goals may require that these persons be linked back to the family at their permanent address, and hence that data be collected for away-from-home family members at their temporary addresses. Data collection costs are generally higher for these away-from-home persons. The alternative would be to consider these away-from-home persons as separate families. For some analyses, it may be appropriate to consider away-from-home persons such as college students as separate families. Studies of poverty and its effect on actions and attitudes would be contaminated by such an approach, however, since the college student "families" would tend to be classified as poor based upon the college student's income.

An additional complicating factor for longitudinal surveys is the changing structure of the family. Throughout the reference period of the survey, families change their composition and new families are created by life events such as birth, death, marriage, divorce, and separation. Defining what constitutes a family becomes much more difficult. To cite a common event, suppose an initial family of mother, father, and child at the start of the survey has the child marrying and leaving home, so that the mother-father-child family becomes a mother-father family and a child-spouse family. Should the mother-father family after the child's marriage be considered the same family as the mother-father-child family or should the mother-father family be considered to be a new family?

The definitional problems that arise in collecting family-level data over time are obviously complex. A simple approach is to define family units based upon living arrangements at the beginning of the reference period and ignore any changes in family structure that occur later. Depending upon survey goals, ignoring primary life events such as marriage, divorce, and separation may prove to be unappealing. Rather, a dynamic definition of the family that recognizes that families can change with time may be more appropriate.

The NMCES adopted a dynamic approach for collecting family-level data, which defined family-level reporting units in each data collection round. After data collection was completed, detailed rules were developed to characterize family unit continuation, dissolution, or formation, based upon these reporting units and the changes that occurred to them over time. In this chapter, the approach that NMCES used for family unit definition and analysis is presented. A comparison is made of families that remain stable over the reference period (stable families) as opposed to families that were created, dissolved, or otherwise changed (dynamic families). Demographic profiles are compared for the two family groups as well as differentials in health care utilization, expenditures, insurance coverage, and disability days (Cohen, 1982).

14.1 The Reporting Unit Data Collection Model

The NMCES data collection strategy was motivated by the analytic goal of measuring health care utilization and expenditures, at individual and family levels, for the civilian, noninstitutionalized population of the United States in 1977. To satisfy this goal, the sampling and data collection plan had to result in data for (1) a probability sample of all persons who were civilian, noninstitutionalized residents of the United States for all or part of 1977, and (2) a probability sample of all families residing in the United States during all or part of 1977 that contained at least one civilian, noninstitutionalized person. To obtain accurate probability samples of these two groups required the development of inclusion rules to account for the many ways persons and families could enter and leave the target population during the reference period of the survey.

The probability sample of persons was obtained in the following manner. First, a multi-stage clustered sample of dwelling units was selected. All civilian, noninstitutionalized residents of these sample dwelling units at the time of the initial Round 1 interview were included in the NMCES. In addition, persons entering the eligible population by birth or return from an institution, the military, or overseas residence were included in the NMCES when they joined a family containing one or more sample persons. The two groups of sample persons form the basis for all person-level analyses. These sample persons are referred to as "key" individuals. Data for these sample persons were gathered for 1977 in six interview rounds. Sample individuals who moved were interviewed at their new location.

For use in the data collection process, a family was defined to be individuals related by blood, marriage, adoption, or foster parent relationship. Unrelated individuals were treated as distinct single member families. These unrelated individuals include single person households, single persons rooming together, lodgers unrelated to the primary family occupants of a dwelling unit, and residents of group quarters.*

The probability sample of families was obtained by including a family and all its members in the NMCES when it contained one or more key individuals. As the reference period progressed, persons were added to the survey because they married a key sample person or went to live with a relative who was a key sample person. Except for the persons joining key person's families that were newborns or returning from the military, overseas residence, or institutions, the persons added to families after Round 1 were classified as nonkey. For nonkey family members, data were collected only for the time period in 1977 when they belonged to a family containing a key individual. The data collected for these nonkey persons are used only in constructing data aggregations needed for family-level analyses (e.g., the annual expenditures incurred by the family for health care in 1977).

At the commencement of the NMCES, it was decided that college students who were unmarried and age 17 to 22 should be considered to be part of their parents' family. These individuals were linked to their parents' residence and only included in the survey when their parents' residence was sampled. Collecting data for college students living away from home presented a problem since data collection

---

*This definition of the family differs from that used by the Census Bureau which does not recognize one person families.

for the student occurred at the temporary college address which was not always within the range of the interviewer assigned to complete the parents' interview. The questionnaire used in the study was set up as a family-level booklet with separate columns on each page for the responses of each family member. Since it was not feasible to ship the questionnaire back and forth between addresses, separate questionnaire booklets were completed for the college students living away from home.

For use in monitoring the status of data collection, reporting units were created. A _reporting unit_ was defined to be individuals residing in the same dwelling unit who are related by blood, marriage, adoption, or foster parent relationship. A family that had a college student member living away from home would have two questionnaire booklets completed and two reporting units established - one for family members interviewed at the parents' address and one for the college student.

For students who returned home (e.g., permanently or for the summer vacation), a separate interview was unnecessary. In these situations, the student was reassigned to his parents' reporting unit and the old reporting unit dissolved. Similarly, when a family member left home to go away to college (e.g., incoming freshmen or students returning from summer vacations), a new reporting unit was established so that data collection could take place at the new location.

The reporting unit data collection scheme, entwined in a panel design, provided an initial framework from which to formally define family units and develop a family unit analysis file. Detailed rules were needed to characterize family continuation, dissolution, or formation in terms of these reporting unit specifications. To monitor the formation and dissolution of families in NMCES, the survey instruments obtained information in each round of data collection concerning change in family composition since the previous interview. One question elicited information on individuals no longer in the original reporting unit, the reasons for the change, and the date on which the move occurred. The new address was obtained for those members who moved, a new reporting unit was assigned to all members of the original household that moved to the new address, and an attempt was made to establish contact. Another question probed to ascertain any additions to the original reporting units.

The process of reporting unit "splits" is a function of movement out of one or more persons, where the spawned family is assigned a new reporting unit number. More specifically, consider an original reporting unit consisting of a household head, spouse, and child (all key participants) in which the child marries and moves out after the initial round of data collection. In the process, the original reporting unit continues its existence and interviews are attempted for the family members. A new reporting unit is created for interviewing the child and spouse at the new address.

14.2 Characteristics of the Reporting Units

In the NMCES, there were 13,955 initial reporting units identified in the first round of data collection for which questionnaire responses were obtained. Over all rounds of data collection, 14,789

Table 14-1. Reporting Unit Distribution

Reporting Unit Characteristics	Number	Percent
Reporting Units Existing for All Rounds of Data Collection	13,514	91.4
Reporting Units with Unchanged Composition for All Rounds of Data Collection	11,653	78.8
Reporting Units with a Change in Composition That Had at Least One Participant with No Change Across All Rounds of Data Collection	1,861	12.6
Reporting Units Existing for Only Part of the Year (Either Created or Dissolved During Study)	1,275	8.6
Total Reporting Units in Existence During Some Period of the Study	14,789	100.0

unique reporting units were identified (Table 14-1). This constitutes a six percent increase in the number of reporting units. There were 11,653 reporting units with the same composition of persons for all rounds of data collection. These unchanged reporting units accounted for 79 percent of all reporting units or 86 percent of the reporting units existing for all five rounds. Of the 13,514 reporting units that existed for all five rounds of data collection, 14 percent had a change in composition. Another 1,275 reporting units were in existence for only part of the year, due to creations or dissolutions that occurred during the study.

The distribution of reporting units by patterns of persons added or persons lost is shown in Table 14-2. There were 1,114

Table 14-2. Distribution of Reporting Units by Patterns of Loss and/or Addition of Members

Reporting Unit Description	Number	Percent
Reporting Units with the Same Composition of Persons Throughout the Survey	11,653	78.8
Reporting Units with Loss of One or More Members During the Survey	1,114	7.5
Reporting Units with the Addition of One or More Members During the Survey	1,774	12.0
Reporting Units with Both Loss and Addition of Members During the Survey	248	1.7
Total Reporting Units	14,789	100.0

reporting units with family members lost across rounds, 1,774 reporting units with family members added across rounds, and 248 reporting units that manifested both additions and losses over the year. As indicated, 79 percent of all reporting units experienced no change in composition. Of the 40,323 NMCES participants, 39,196 belonged to only one reporting unit over all rounds of data collection. There were another 1,061 participants who belonged to two distinct reporting units, and 66 participants with links to three reporting units over the year.

The reporting unit structure allows for the derivation of national estimates of health care measures and demographic characteristics at the family level, where family definition is consistent with the definitions that identify reporting units. For the purpose of estimation, reporting unit sampling weights were defined in terms of the reciprocal of their selection probabilities. Since the reporting units owe their creation to their linkage with the dwelling units selected for inclusion in the NMCES, they retain the dwelling unit's probability of selection. Reporting units that were formed by splits of an existing reporting unit retain the sampling weight of the original unit. In those rare circumstances when two reporting units merge, the average of their sampling weights can be used to characterize the weight of the newly formed unit.

14.3 Definition of Family Units

The restrictive definitions of families that were implicit in the reporting unit model were inappropriate for satisfying the analytical goals of NMCES family analyses. Since many of the proposed analyses focused on the health utilization, expenditure, and insurance coverage data at the family level, which were assumed to be sensitive to changes in family composition, an alternative family unit analysis strategy was required. The family unit definition coincided with the reporting unit specifications, in terms of consisting of one or more individuals related by blood, marriage or adoption, but also included college students temporarily living away from home. A family was said to have changed and to be a different family when the head or spouse moved out.*

---

*Alternative definitions of family unit formation and dissolution, motivated by specific analytical concerns, have also been explored in the NMCUES (Dicker and Casady, 1982).

The difference in family unit and reporting unit definitions was manifest in the treatment of the formation and dissolutions of families. In defining reporting units, separation and divorce usually resulted in the person(s) moving out being assigned a new reporting unit number and the person(s) that remained at the old address keeping the same reporting unit number. In the family unit framework, two new families were formed when the head or spouse moved out of an existing family unit and the original spawning family ceased to exist. Whenever there was a loss of the head or spouse due to death, institutionalization or movement into military, a new family was formed and, as before, the original family ceased to exist. For changes in family composition concerning family members other than head or spouse, such as birth, death, movement out, or a member institutionalized or joining the military, the family was considered to be the same family, albeit with a different number of members.

Following the rules outlined above, the NMCES family units were constructed from the reporting units defined during data collection. The composition of most family units were unaffected by change during the survey year, as would be expected from trends in the general population. Those families that exhibited transformations were properly accounted for to allow the generation of dynamic estimates of family unit health experiences over the survey year.

14.4 The Family Unit Weighting Strategy

The derivation of national estimates of health care parameters for families in a panel design requires the appropriate formulation of analyses weights. The population of family units existing on any one day during the calendar year 1977 is potentially different from that existing on any other day due to the formation and dissolution of families occurring over time. The families are defined with a beginning date, an ending date and the set of participants in the household survey that are members of the family. Horvitz and Folsom (1980) outline an analysis approach that exploits the linkage between potential Round 1 NMCES survey participants and the universe of all family units spawned by these persons in 1977. The strategy as it was implemented for the NMCES has been described by Bentley and Folsom (1981) and will be summarized in the remainder of this section.

To characterize this linkage between unique families and potential Round 1 participants, family membership indicators need to be defined. Let $g(ij)$ be defined as a zero-one indicator such that

$$g(ij) = \begin{cases} 1 & \text{if the i-th participant ever belonged to family unit j in 1977, and} \\ 0 & \text{otherwise} \end{cases}$$

The total number of potential Round 1 NMCES participants belonging to family unit j is defined as the family unit's multiplicity and is specified in terms of the membership indicators as $g(+j)$, where

$$g(+j) = \sum_{i=1}^{I} g(ij) \tag{14-1}$$

and I is the total number of potential Round 1 NMCES survey participants, defined as the civilian noninstitutionalized population during Round 1.

To determine the total number of family units existing on day t, existence indicators $E_t(j)$ for time t and family j must be defined, where

$$E_t(j) = \begin{cases} 1 \text{ if family unit j exists on day t} \\ 0 \text{ otherwise} \end{cases}$$

Consequently, the sum of these existence indicators over all J family units, or

$$F(t) = \sum_{j=1}^{J} E_t(j) \qquad (14\text{-}2)$$

provides a count of the total number of family units existing on day t. This can be re-expressed using a multiplicity framework (Sirken, 1970) such that

$$F(t) = \sum_{i=1}^{I} \sum_{j=1}^{J} g(ij)\, E_t(j)/g(+j) \qquad (14\text{-}3)$$

This expression is now defined in terms of participant-level quantities. Since the NMCES family units are distinguished by their composition of sample participants, the sample weight for a family can be formulated to incorporate the weights of its members. The individual's unadjusted sampling weight is the reciprocal of its selection probability. This selection probability is equivalent to the probability of selection for the sample dwelling unit within which the sample participant resided in Round 1. These sampling weights were further adjusted to reflect the nonresponse of entire reporting units and smoothed to reduce the effect of unequal weighting (Cohen and Kalsbeek, 1981).

Within this framework, an unbiased estimate of the number of family units existing on day t, F(t), can be derived as

$$\hat{F}(t) = \sum_{i \varepsilon S} W(i) \sum_{j=1}^{J'} g(ij)\, E_t(j) \,/\, g(+j) \tag{14-4}$$

where W(i) denotes the nonresponse-adjusted weight for the Round 1 sampling unit that contained participant i, $i \varepsilon S$ indicates all eligible participants listed in the rosters of Round 1 reporting units, and J' represents all sampled families.

In this setting, the family unit weights are defined as

$$W_f(j) = \sum_{i \varepsilon S} W(i)\, g(ij) \,/\, g(+j) = \sum_{i \varepsilon S(j)} W(i) \,/\, g(+j) \tag{14-5}$$

where S(j) includes all eligible household participants found in responding Round 1 reporting units, who at one time during 1977 belonged to family unit j.

These weights were further adjusted to provide annual and quarterly estimates of the number of families existing during 1977. Their application allows for the estimation of family health care utilization and expenditure characteristics at the national level.

14.5 A Comparison of Demographic Measures Between Stable and Dynamic Families

A critical analytical concern of the family-level data analysis was the determination of those characteristics that distinguished families experiencing a change in composition over the survey year and those remaining intact. Consequently, a comparison of demographic measures that characterize these distinct populations was considered. From NMCES family-level data, it was possible to classify the sample of family units into two mutually exclusive groups identifying stable and dynamic family types. More specifically, the stable group consisted of families with no change in membership over the year, but allowing for the nonresponse of individual members. The dynamic group consisted of families with at least one member belonging to more than one family. In the NMCES survey, 86 percent of the families were classified as stable.

The demographic measurements on families were based upon Census region and size of city and the following characteristics of the family head: age, education level, employment status, health status, martial status, poverty status, sex, and race. Parameter estimates of the demographic distributions for the alternative family types can be observed in Table 14-3.

To determine whether the estimated parameters of the distributions were equivalent across family types, a test of homogeneity appropriate for data from a complex survey design was considered. First the vectors of estimated proportions characterizing each demo-

Table 14-3. A Comparison of Demographic Measures Between Stable and Dynamic NMCES Families

Demographic Measures	Stable Families Proportion	Stable Families Standard Error	Dynamic Families Proportion	Dynamic Families Standard Error
Age of Head:				
0 to 18	0.001	0.001	0.012	0.002
19 to 24	0.061	0.003	0.150	0.009
25 to 54	0.538	0.007	0.537	0.020
55 to 64	0.170	0.004	0.167	0.017
65 or over	0.230	0.007	0.133	0.012
Q = 112.243 (4 degrees of freedom), P < 0.001*				
Years of Education:				
0 to 8	0.224	0.008	0.226	0.017
9 to 11	0.160	0.004	0.155	0.014
12	0.315	0.006	0.317	0.018
13 to 15	0.137	0.004	0.157	0.016
16 to 18	0.164	0.007	0.145	0.015
Q = 1.611 (4 degrees of freedom), P = 0.807*				

Employment Status:				
Worked	0.660	0.007	0.695	0.017
Unemployed	0.022	0.002	0.029	0.004
Not in Labor Force	0.317	0.007	0.275	0.016
Q = 6.074 (2 degrees of freedom), P = 0.048*				
Health Status:				
Excellent	0.429	0.007	0.406	0.018
Good	0.378	0.006	0.382	0.016
Fair	0.143	0.005	0.155	0.011
Poor	0.051	0.003	0.057	0.006
Q = 2.459 (3 degrees of freedom), P = 0.483*				
Marital Status:				
Never married	0.101	0.005	0.162	0.013
Married	0.641	0.007	0.611	0.016
Widowed	0.143	0.004	0.084	0.009
Separated	0.037	0.003	0.046	0.007
Divorced	0.077	0.003	0.095	0.009
Q = 59.414 (4 degrees of freedom), P < 0.001*				

Table 14-3. (continued)

Demographic Measures	Stable Families		Dynamic Families	
	Proportion	Standard Error	Proportion	Standard Error
Poverty Status:				
Poor, Below Poverty Level	0.123	0.004	0.125	0.010
Near Poverty Level	0.051	0.002	0.039	0.006
Low Income	0.142	0.004	0.140	0.011
Middle Income	0.351	0.005	0.335	0.014
High Income	0.333	0.008	0.362	0.017
Q = 4.796 (4 degrees of freedom), P = 0.3089*				
Census Region:				
Northeast	0.221	0.015	0.185	0.024
North Central	0.262	0.017	0.290	0.026
South	0.315	0.024	0.334	0.030
West	0.202	0.012	0.191	0.016
Q = 2.675 (3 degrees of freedom), P = 0.445*				

Size of City:				
SMSA: 16 largest SMSAs	0.257	0.021	0.232	0.030
SMSA: Population 500,000+ but not 16 largest SMSAs	0.263	0.030	0.274	0.036
SMSA: Population less than 500,000	0.180	0.031	0.188	0.035
Not SMSA: less than 60% rural	0.181	0.022	0.188	0.026
Not SMSA: 60% or more rural	0.118	0.020	0.118	0.022
Q = 0.593 (4 degrees of freedom), P = 0.964*				
Sex:				
Male	0.749	0.006	0.702	0.012
Female	0.251	0.006	0.298	0.012
Q = 12.272 (1 degree of freedom), P = 0.0005*				
Race:				
White	0.882	0.008	0.874	0.017
Nonwhite	0.118	0.008	0.126	0.017
Q = 0.181 (1 degree of freedom), P = 0.670*				

*Testing for homogeneity in distributions across family types involve determining the P-value for the Wald Statistic, Q, by comparison to a χ^2 distribution with the appropriate degrees of freedom.

graphic distribution and their associated variance-covariance matrices were specified for the stable and dynamic families. Variances and covariances of all the estimated population parameters were derived using the Taylor Series linearization method (Woodruff, 1971; Shah, 1979). Weighted least squares methodology was then used to test for the equivalence of parameter estimates across family types.*

The age of head distributions for the stable and dynamic families differed significantly at the 0.05 level (Table 14-3). The difference was most notable across family types within age of head categories 19 to 24 and 65 or over. The stable families exhibited a significantly greater proportion of family heads 65 or over than in the dynamic group. This supports the notion that the stable families have a greater concentration of older families than their dynamic counterparts. In contrast, the dynamic families had a greater representation of family heads in the younger age categories. These families are composed of younger individuals more likely to experience the effects of marriage, divorce, graduation from college, or assumption of new employment, and thus form new family units.

The marital status distribution that characterized family heads also differed significantly across family types. This differential was primarily explained by the greater proportionate representation of never married family heads characterizing the dynamic set, as

---

*The weighted least squares methodology has been described by Grizzle, et al (1969) and implemented in the analysis of data from complex surveys by Koch, et al (1975) and Freeman, et al (1976). The methodology is described in greater detail in Chapter 11.

contrasted by the larger relative frequency of families with widowed heads in the stable group. In addition, a significantly greater representation of families with female heads was observed in the dynamic group.

It was also noted that the distribution of the employment status of the family head differed significantly between the stable and dynamic families. The stable families exhibited a greater proportional representation of family heads not in the labor force. Again, these families are more likely to have older heads. Comparisons of the population distributions characterized by Census region and size of city measures revealed no significant differences across family types. Consequently, there was no observed disproportionate concentration of either family group in a particular geographic setting, whether distinguished by region or metropolitan and non-metropolitan breakdowns. Similarly, no significant differences between the stable and dynamic families were noted when examining distributions characterized by the family head's years of education, health status, poverty status, and race.

14.6 An Examination of Health Care Differentials Between Members of Stable and Dynamic Families

In the NMCES, 91.2 percent of the key survey participants were members of stable families. The observation of significant differentials in the demographic profiles for the stable and dynamic families motivated additional analyses at the individual level. These analyses were initiated to determine whether a stable or dynamic family unit classification was associated with significant differentials in the health care behavior of individuals. The observation of significant differences would suggest the presence of a family dynamism/stability effect.

To provide for a comprehensive investigation, the health care measures investigated included annual data on utilization, expenditures, disability and health insurance coverage for 1977. The health care utilization measures under consideration included the number of physician visits, hospital admissions, and prescribed medicine use. More specifically, physician visits consisted of all ambulatory physician contacts, excluding telephone calls. Hospital admissions included admissions of less than 24 hours and those for women giving birth. Newborn were not counted separately following delivery. Prescribed medicines included any drug or other medical preparation prescribed by a physician, including refills.

Expenditure data for each of these utilization measures were also considered: physician visit expenditures, total expenditures for prescribed medicines, and total expenditures for all hospital admissions (excluding charges for inpatient physician services). The measure of disability that was of interest consisted of the number of days that illness or injury kept an individual in bed, away from a job, other work, or usual activity (e.g., work around the house, school). Measures of health insurance coverage included indicators for the presence of private coverage at some time during the survey year, public or private coverage for the entire year, and lack of any coverage for all of 1977.

Parameter estimates and related standard errors of the health care measures for members of stable and dynamic families can be observed in Table 14-4. All tests of statistical significance were performed at the 0.05 level and based upon the asymptotic normality of the Studentised differences.

The mean number of hospital admissions and related expenditures differed significantly for the members of stable and dynamic families. The hospital utilization and expenditure experience was significantly higher for members of dynamic families. The mean number of prescribed medicines and related expenditures were also observed to differ significantly across the alternative family types. However, the prescribed medicine utilization and expenditure pattern was significally higher for members of stable families. No significant differentials were noted for the mean number of physician visits and related expenditures. Similarly, the mean number of disability days experienced by members of the alternative family types were determined to be statistically equivalent.

Examination of the respective health insurance coverage profiles also revealed a significant differential. Members of stable families were more likely to possess some form of health insurance coverage for the entire year. This difference was anticipated, given the potential for change in coverage for individuals who were members of more than one family over the survey year. A statistically equivalent proportion of the members of the alternative family types were without health insurance coverage for all of 1977. Similarly, no significant differentials were observed between family types in the percent ever covered by private insurance.

The observation of significant health care differentials for members of the alternative family types demonstrated a need for

Table 14-4. A Comparison of Health Care Measures For Members of Stable and Dynamic NMCES Families

Health Care Measure	Members of Stable Families		Members of Dynamic Families		Studentised Difference
	Estimate	Standard Error	Estimate	Standard Error	
Mean Number of Physician Visits	4.02	0.05	3.93	0.15	0.57
Mean Number of Hospital Admissions	0.15	0.01	0.18	0.01	-2.87*
Mean Number of Prescribed Medicines	4.37	0.10	3.88	0.22	2.03*
Mean Expenditures for Physician Visits	99.26	1.80	102.24	4.68	-0.59

Mean Expenditures for Hospital Admissions	209.20	7.48	280.00	30.18	-2.28*
Mean Expenditures for Prescribed Medicines	26.52	0.68	23.39	1.40	2.01*
Mean Number of Disability Days	13.06	0.40	15.04	1.01	-1.82
Percent Ever Covered by Private Insurance	78.49	0.83	75.38	1.45	1.86
Percent Always Covered by Private or Public Insurance	84.42	0.53	76.01	1.27	6.11*
Percent Always Uninsured	8.52	0.44	10.5	0.98	-1.84

*Indicates a statistically significant difference at the 0.05 level.

additional analyses that directly tested for the presence of a family dynamism/stability effect. This motivated an investigation to determine whether the observed health care differentials were driven by this effect, and not just a consequence of differences in associated predispositional factors. Consequently, it was necessary to control for predispositional factors, in order to isolate the hypothesized family dynamism/stability effect. The predispositional factors that were considered, consisted of those measures identified in the demographic comparisons. The respective classes for each of the predispositional factors were generally consistent with their previous specification, although they now represented the respective family member's profile. The following measures were redefined: health status (excellent or good, fair or poor), poverty level (poor, near poor or low income, middle income, high income), size of city (SMSA, nonSMSA), and education (0 to 8, 9 to 11, 12, 13 to 15, 16 or more, under 17 years of age). An insurance coverage measure (some coverage, never covered) was included in the analysis of the utilization and expenditure measures.

For the utilization and expenditure measures exhibiting significant differentials across family types in the univariate comparisons, a regression strategy was applied to test for the dynamism/stability effect. The approach considered a weighted least squares methodology, appropriate for complex survey data, where variances of estimated model parameters were derived using the Taylor Series linearization method (Holt, et al, 1980). A logistic regression analysis was adopted for the insurance measure indicating coverage for the entire year (Nerlove and Press, 1973). In this setting, a design effect adjustment was applied to the variances of estimated model parameters, to accomodate the complex survey design (Cohen, 1983).

A significant family dynamism/stability effect was noted with respect to hospitalizations and related expenditures (Table 14-5). Consequently, membership in a dynamic family was associated with a higher utilization and expenditure pattern for hospital admissions, even when controlling for the effects of related predispositional

Table 14-5. Estimated Family Dynamism/Stability Effect*

Health Care Measure	Estimated Parameter	Standard Error	Partial-F	P Value
Hospital Admissions	- 0.037	0.013	8.12	<0.01
Hospital Expenditures	-99.445	30.935	10.33	<0.01
Prescribed Medicine Utilization	0.121	0.195	0.39	0.54
Prescribed Medicine Expenditures	0.396	1.206	0.11	0.74
Possession of Health Insurance Coverage for All of 1977	0.337	0.076	19.82	<0.01

*A multivariate regression was used for the utilization and expenditure measures, and a logistic regression analysis was applied to the insurance coverage data. The partial F-Test determines the presence of a family dynamism/stability effect on health care behavior, after controlling for specified predispositional factors. The indicator variable for family membership was coded X=1 for members of stable families and X=0 for members of dynamic families.

factors. A similar outcome was noted for the logistic analysis of the insurance coverage measure. After controlling for the specified predispositional factors, membership in a stable family was still characterized by a greater likelihood of possessing coverage for all of 1977. The significant differentials previously observed for prescribed medicine data were primarily due to differences in the predispositional factors that characterized members of the alternative family types. Once these factors were controlled for, the family type classification exerted no additional influence on health care behavior. Consequently, it was determined that the presence of a family dynamism/stability effect was selective in nature, and was often non-operational.

14.7 Summary

In this study, the framework for NMCES family unit analysis was presented, which provides for the derivation of national estimates of family health characteristics. The weighting strategy uses a multiplicity framework to yield annual and quarterly estimates of the average number of families existing during the period, in addition to national estimates of relevant health care measures.

An analysis to determine demographic characteristics that distinguish stable and dynamic families was also presented. Findings revealed significant differences across families by age, marital status, sex, and employment status of the family head. In addition, the health care behavior for members of the alternative family types was examined. Significant differences were noted for prescribed medicine and hospital use, related expenditures, and for a measure of health insurance coverage. When differentials in predispositional factors associated with the health care indices were controlled, however, a family dynamism/stability effect was not typical.

References

Bentley, B. S., and R. E. Folsom (1981). Family Unit Analysis Weighting Methodology for the National Medical Care Expenditure Survey, RTI Final Report No. 1320-11F, Contract No. HRA 230-76-0268, Available from the National Center for Health Services Research, Rockville, Maryland.

Cohen, S. B. (1983). Design Effect Variation in the National Medical Care Expenditure Survey. Proceedings of the American Statistical Association, Survey Research Methods Section, 748-753.

Cohen, S. B. (1982). Family Unit Analysis in the National Medical Care Expenditure Survey. Proceedings of the American Statistical Association, Survey Research Methods Section, 561-566.

Cohen, S. B. and W. D. Kalsbeek (1981). NMCES Estimation and Sampling Variances in the Household Survey, Instruments and Procedures 2, DHHS Publication No. (PHS) 81-3281, Washington, DC: U.S. Government Printing Office.

Dicker, M. and R. J. Casady (1982). A Reciprocal Rule Model for Defining Longitudinal Families for the Analysis of Panel Survey Data. Proceedings of the American Statistical Association, Social Statistics Section, 532-537.

Freeman, D. H. Jr., J. L. Freeman, D. B. Brock, and G. G. Koch (1976). Strategies in the Multivariate Analysis of Data from Complex Surveys II: An Application to the United States National Health Interview Survey. International Statistical Review, 44, 317-330.

Grizzle, J. E., C. F. Starmer, and G. G. Koch (1969). Analysis of Categorical Data by Linear Models. Biometrics, 25, 489-504.

Holt, D., T. M. F. Smith, and P. D. Winter (1980). Regression Analysis of Data from Complex Surveys. Journal of the Royal Statistical Society, 143(4), 474-487.

Horvitz, D. G. and R. E. Folsom Jr. (1980). Methodological Issues in Medical Care Expenditure Surveys. Proceedings of the American Statistical Association, Survey Research Methods Section, 21-29.

Koch, G. G., D. H. Freeman Jr., and J. L. Freeman (1975). Strategies in the Multivariate Analysis of Data from Complex Surveys. International Statistical Review, 43, 59-78.

Nerlove, M. and S. J. Press (1973). Univariate and Multivariate Log-Linear and Logistic Models. Rand Report No. R1306EDA/NIH, Available from Rand Corporation, Santa Monica, California.

Shah, B. V. (1979). SESUDAAN: Standard Errors Program for Computing of Standardized Rates from Sample Survey Data, RTI Report, Research Triangle Park, North Carolina.

Sirken, M. G. (1970). Household Surveys With Multiplicity. Journal of the American Statistical Association, 65, 257-266.

Woodruff, R. S. (1971). A Simple Method for Approximating the Variance of a Complicated Estimate. Journal of the American Statistical Association, 66, 411-414.

15
A Comparative Study of Synthetic Estimation Procedures

Planners, researchers, government and the business sector are a few of many sources that require reliable statistics to formulate and implement policies at the local and state levels. Due to cost and sample size considerations, state and local estimates of health, social, political and economic parameters are not directly available from most national surveys. To satisfy the need for subnational estimates, several alternative strategies have been developed that use national survey data to construct subnational estimates. These techniques develop prediction models using available survey data at the national level, data on population characteristics for the subnational area of interest, and auxiliary data related to the variable of interest. Unlike direct estimators with sufficient sample data from every area for which separate estimates are desired, synthetic estimators are model-based estimators that use auxiliary information from small areas together with national survey data to develop a small area estimate.

The variable for which an estimate is required is commonly referred to as the <u>criterion variable</u> and the subnational areas for which estimates are needed are referred to as <u>small areas</u>. Synthetic estimation is commonly referred to as <u>small area estimation</u> since estimates are synthetically developed for areas too small - in terms of a national survey - to warrant the sample size needed to produce direct estimates. To construct synthetic estimates, data are needed about the characteristics of the small area. These data may include demographic data from the last Census together with population counts obtained from government records. Information that can be used to predict the value of the criterion variable for the small area is particularly useful. This information is referred to as <u>symptomatic information</u> since the information is indicative of the value of the criterion variable. As an example, an estimate might be desired of the total expenditures for medical care in 1977 for residents of a particular state. The criterion variable then is 1977 expenditures for medical care. At the state-level, population counts might be available for categories defined by age, race, and sex. Symptomatic data that might also be available would include percent on Medicaid, the number of hospital beds in the state, and the percent aged. Finally national survey data from the 1977 NMCES could be used, together with this demographic and symptomatic data, to calculate the synthetic estimate for the state.

As a consequence of empirical tests of validity and widespread usage, a limited number of these procedures have gained respectability under certain qualifying conditions. No technique is consistently superior in accuracy to its alternatives. The type of predictor information available, the functional relationship of the predictor information to the specified criterion variables, and the assumptions underlying each prediction model narrow the number of procedures that are appropriate for a particular application. The results of previous empirical studies should also be considered when they involve related criterion variables and comparisons were made of the precision of alternative techniques.

The small area estimation procedures whose use is most widespread in the statistical literature include:

- the NCHS synthetic estimator,
- the sample-regression estimator,
- the post-stratified estimator, and
- the composite estimator.

In this chapter, these small area estimation strategies are described and their applicability for data from the National Medical Care Expenditure Survey is examined (Cohen, 1980). The NMCES data base is particularly attractive in its similarity of structure to many of the national survey data bases used in developing synthetic subnational estimates.

15.1 The NCHS Synthetic Estimator

One of the first small area estimators was developed by the National Center for Health Statistics (NCHS). The *NCHS synthetic estimator* is based upon the assumption that, within demographic groups, the characteristics of the population of the small area of interest is similar to that of the nation or the geographic region in which the area is located (National Center for Health Statistics, 1968). The NCHS estimator requires that demographic data (e.g., population counts by race, income, sex, age) be available for the small area that are related to the criterion variable being estimated. National survey data are then used to develop estimates of the criterion variable for each of the D domains defined by cross-classification of the demographic variables.

Suppose that an estimate of the small area ℓ mean for a criterion variable Y is desired. To produce the synthetic estimate of the mean $\bar{Y}(\ell)$, the NCHS model computes a weighted average of the survey-estimated domain means, with weights that reflect the proportional distribution of the D domains in the small area ℓ population. That is,

$$\bar{y}_n^*(\ell) = \sum_{d=1}^{D} P(\ell d)\ \bar{y}(d) \tag{15-1}$$

where

$\bar{y}_n^*(\ell)$ is the NCHS synthetic estimator of the mean for the criterion variable in small area ℓ,

$P(\ell d)$ is the proportion of the ℓ-th area's population belonging to domain d, and

$\bar{y}(d)$ is a national survey estimate of the mean value of the criterion variable for domain d.

More detailed equations for the NCHS synthetic estimator include a regional adjustment.

This procedure is particularly appealing as a consequence of its straightforward application to survey data. However, NCHS synthetic small area estimates generally cluster near the mean for the nation or the specific geographical region (when a regional adjustment is used). Consequently, the method is not particularly sensitive to internal forces operating within the small area. By assuming the small areas share the same characteristics as the nation within demographic groups, small area differences can only be distinguished by differences in the demographic configuration of the area. In addition, unbiased estimates are required of the proportion of the population of the small area found in each domain. This condition is easier to satisfy when small area estimates are required in a census year. The availability and accuracy of direct estimates for the $P(\ell d)$ quantities diminishes as the time since the last Census increases.

15.2 The Sample-Regression Estimator

Ericksen (1974) described an alternative technique for computing small area estimates which combines symptomatic information and sample data into a multiple regression format (assuming an underlying linear model). Referred to as the Ericksen sample-regression method of small area estimation, the method assumes the availability of criterion variable estimates for a sample of small areas (e.g. counties) and symptomatic indicators for the universe of small areas.

The _sample-regression estimation procedure_ may be outlined as follows. Initially, criterion variable data must be available for a sample of n small areas selected from the set of N small areas in the total population. These small areas are referred to as primary sampling units (PSUs) and the n small areas in the sample as sampled PSUs. Estimates of the criterion variable are computed for these sampled PSUs. Symptomatic information is also needed for the sampled PSUs and the nonsampled PSUs for which synthetic estimates are to be made. Using data for the sampled PSUs only, a regression equation is developed to predict the criterion measure based upon the symptomatic information. The regression model that is fit to the criterion variables estimates takes the matrix representation:

$$\bar{Y} = X\beta + E \tag{15-2}$$

where

$\bar{Y}$ is an n by 1 vector of values for the criterion variable in the n sampled PSUs,

X is an n by (p+1) matrix containing the set of p symptomatic indicators for the n sampled PSUs (and including an indicator for an intercept term),

β is a (p+1) by 1 vector of regression coefficients, and

E is an n by 1 vector of stochastic errors.

Estimates for small areas are then determined by substituting values of the symptomatic indicators into the estimated regression equation. That is, the sample-regression estimator for small area ℓ is obtained as

$$\bar{y}_r^*(\ell) = X(\ell)\,\hat{\beta} \tag{15-3}$$

where $X(\ell)$ is the (p+1) by 1 vector of symptomatic information for local area ℓ and $\hat{\beta}$ is the regression estimate obtained in fitting the model in equation (15-2) to the data from the sampled PSUs.

15.3 The Post-Stratified Synthetic Estimator

The sample-regression method advanced by Ericksen is most appropriate when the linearity assumption is satisfied and the multiple correlation for the regression model is reasonably large. For those situations when the multiple correlation level is moderate (0.4 to 0.7) and a nonlinear model is more appropriate, Kalsbeek (1973) and Cohen and Kalsbeek (1977) have described a post-stratified synthetic estimation procedure whose most limiting assumption is the availability of symptomatic information. Generally, synthetic estimation procedures treat the small areas as the lowest level for which estimates are made. In contrast, the Kalsbeek-Cohen post-stratified synthetic estimator breaks up the small area into constituent geographical sectors or "base units," which might be counties, enumeration districts, or other geographical subunits of a county. The small area for which a variable of interest is to be estimated is referred to as the "target area" and further subdivided into "target area base units." Unlike other methods that use symptomatic information directly for estimation, this procedure uses the symptomatic information to group base units (sample base units) from the total population.

The <u>post-stratified synthetic estimator</u> may be defined as follows. Again, survey data for a sample of n base units are needed from the total population of N base units. Both symptomatic and criterion information are needed for the sample base units. These base units are divided into S groups or <u>post-strata</u> based upon the symptomatic information only. The object is to form post-strata that are internally homogeneous with respect to the criterion variable while dissimilar from one another. Grouping can be handled by any one of several hierarchical procedures in cluster analysis (Johnson, 1967; Anderberg, 1973; Everitt, 1980) or by minimum variance stratification schemes (cum $\sqrt{f}$ rule of Dalenius and Hodges, 1959; Cochran, 1963). It is noteworthy that the post-strata groups need not be symmetric. Small cells resulting from the cross-classification of several variables may be combined to yield post-strata of sufficient size.

All target area base units belonging to the small area of interest are then assigned (classified) to one of the S post-strata based upon the symptomatic information. An estimate of the criterion variable for each of the target area base units is obtained from the sample base units in the post-stratum to which it has been assigned. In essence, each target area base unit estimate can be perceived as a synthetic estimate. These synthetic estimates are then pooled to arrive at a final estimate for the respective target area. Specifically, the estimate for the s-th post-stratum ($s = 1,2,\ldots,S$) of the criterion variable of interest takes the form:

$$\bar{y}(s) = \sum_{i \varepsilon s} W(i)\, \bar{y}(i) \tag{15-4}$$

where

$W(i)$ estimates the proportion of the total population of sample base units classified in post-stratum s that are represented by base unit i, and

$\bar{y}(i)$ is an estimate of the criterion variable of interest for the i-th sample base unit.

The estimate of the criterion variable for each local area ℓ takes the form:

$$\bar{y}_p^*(\ell) = \sum_{s=1}^{S} P(\ell s)\ \bar{y}(s) \tag{15-5}$$

where $P(\ell s)$ is the proportion of the population of small area ℓ that belongs to post-stratum s.

The post-stratified estimator bears a striking resemblance to the NCHS synthetic estimator previously discussed. The primary difference lies in the method of post-strata construction versus domain formation. For the NCHS method, domain estimates are generated once _individuals_ are classified according to their demographic characteristics. In contrast, the post-stratified estimator links all individual observations within a sample base unit to a particular post-strata, based upon symptomatic information for the unit. Population estimates for these $P(\ell s)$ quantities may be more readily obtainable than the demographic distributions required by the NCHS estimator.

15.4 The Composite Synthetic Estimator

A final method which has recently gained attention is the _composite synthetic estimator_ which takes the form of a weighted average of two component estimators for the small area or

$$\bar{y}_c^*(\ell) = C(\ell)\ \bar{y}_1^*(\ell) + [1 - C(\ell)]\ \bar{y}_2^*(\ell) \tag{15-6}$$

where $C(\ell)$ is an appropriately chosen weight and $\bar{y}_1^*(\ell)$ and $\bar{y}_2^*(\ell)$ are alternative synthetic estimators for local area ℓ. The mean square error of the composite synthetic estimator is smaller than the larger of the mean square errors for the two component estimators. Researchers often do not have a priori information to determine the most accurate synthetic estimator for a criterion variable of interest. The composite estimator is a welcome alternative that minimizes the selection of a less accurate technique.

Schaible, Brock, and Schnack (1977) have demonstrated that with judicious selection of composite weights, the composite estimator can have a mean square error that is smaller than the mean square error of the individual estimators. When $E\{[\bar{y}_1^*(\ell)-\bar{y}(\ell)][\bar{y}_2^*(\ell)-\bar{y}(\ell)]\}$ is small relative to MSE $[\bar{y}_2^*(\ell)]$, the weight that will minimize the composite estimator's mean square error can be approximated as

$$C(\ell) = 1/[1+R(\ell)] \tag{15-7}$$

where

$$R(\ell) = MSE[\bar{y}_1^*(\ell)] \;/\; MSE[\bar{y}_2^*(\ell)] \tag{15-8}$$

15.5 A Comparison of Alternative Synthetic Estimation Strategies

When conducting reliability studies of alternative small area estimation strategies, a major limitation is the lack of independent small area estimates for the criterion variable. Consequently, measures of bias and mean square error are often difficult to directly estimate. However, design features of the National Medical Care Expenditure Survey provide a solution to this constraint for criterion measures of health care usage and expenditures. The complex nature of the survey design, which used two independently-selected national samples, is ideally suited for the evaluation of methodologies used in small area estimation. The replicated property of the survey design yields independent unbiased estimates for two probability samples of local areas and makes possible reliability studies of the quality of alternative synthetic estimates used with NMCES data.

Because of the time span between the 1977 NMCES and the 1970 Census, accurate small area statistics were unavailable for the population totals in the demographic-defined domains used by the NCHS synthetic estimator. Though some marginal distributions of the demographic variables were available for larger areas such as states, it was thought to be undesirable to estimate the required small area population totals and then use these estimates (rather than known population counts) in the synthetic estimation. For this reason, the NCHS synthetic estimator was not investigated. Instead, the sample-regression estimator and the post-stratified estimator and a composite of the two estimators were investigated for use with NMCES data.

NMCES data on insurance coverage were used to derive unbiased PSU-level estimates of the following criterion variables:

- population percentage ever on Medicaid,
- population percentage ever on Medicare,
- population percentage ever covered by private insurance, and
- population percentage never insured in 1977.

Only the NORC half-sample of 76 PSUs were used in the determination of synthetic estimates for the alternative methods. These PSUs served as the sample base units for the post-stratified model and as observational units for the sample-regression approach. The composite estimator was derived as a weighted function of these two estimators. Synthetic estimators were derived for small areas for which the RTI half-sample provided a direct estimate. Measures of the accuracy of the synthetic estimates were then computed using these direct survey estimates.

Stepwise regression was implemented to determine the best symptomatic data to use in predicting the insurance coverage criterion variables from several relevant symptomatic variables abstracted from the _County and City Data Book 1977_ (Bureau of the Census, 1977). For Medicaid coverage, measures of population density, income, and the average number of children were chosen. For Medicare coverage, measures of the aged population, income, and facility supply were selected. Similarly, measures of the aged population, physician supply, and the birth rate were predictors for private insurance. Income, physician supply, and birth rate were predictors for the uninsured. The percent of variation in the criterion variable explained by these predictor variables (R^2) was only moderate (Table 15-1).

With the selection of appropriate symptomatic data, synthetic estimates of the different measures of insurance coverage could easily be computed for the 59 PSUs in the RTI half-sample of the NMCES. Since unbiased estimates of insurance coverage for the 59 RTI PSUs were available from the NMCES sample itself, the bias and mean square error could be estimated for the synthetic estimators.

Table 15-1. Selected Predictor Variables for Each Criterion Variable

Criterion Variable	Symptomatic Variable	R^2
Percent Medicaid	Population per Square Mile Supplemental Security Income Recipients Birth Rate Per 1,000 Persons	0.2581
Percent Medicare	Persons 65 years old and over Hospital Beds Per 100,000 Persons Per Capita Money Income	0.3557
Percent Privately Insured	Persons 65 years old and over Birth Rate Per 1,000 Persons Physicians Per 100,000 Persons	0.2175
Percent Never Insured	Birth Rate Per 1,000 Persons Physicians Per 100,000 Persons Per Capita Money Income	0.3337

The small number of NORC PSUs (76) restricted the number of post-strata formed by the post-stratified estimator to eight to insure an average sample size of nine PSUs for each post-stratum estimator. A minimum variance stratification scheme which employed the cum $\sqrt{f}$ rule on the marginal distributions of the symptomatic variables was used in formation of the post-strata. Previous studies by Cohen (1979) have demonstrated that further increases in the number of post-strata at the expense of fewer sample base units per post-stratum coincide with diminishing returns in terms of reduction of the estimator's mean square error.

The composite estimator was a weighted function of the sample-regression and post-stratified estimators. Two weights were used to construct the composite estimator. One choice approximated the weight that would minimize the composite estimator's mean square error. This weight was defined individually for each area ℓ as $C(\ell)=1/[1+R(\ell)]$. The approach assumes the availability for the criterion variable of unbiased estimates of the mean square errors for the component synthetic estimators. Since the mean square errors for synthetic estimators may not be easily obtainable, a constant weight set at $C(\ell) = \frac{1}{2}$ was also used to form a second version of the composite estimator that averaged the mean square error associated with the two component estimators.

To measure the accuracy of the four small area estimators, the average mean square error was estimated for the synthetic estimates. An unbiased estimator of the mean square error for the synthetic estimate produced by method s for the ℓ-th RTI PSU (ℓ=1,2,...,59) takes the form (Gonzalez and Waksberg, 1973; Gonzalez, 1973):

$$\text{MSE}[\bar{y}_s^*(\ell)] = [\bar{y}_s^*(\ell) - \bar{y}(\ell)]^2 - \text{Var}[\bar{y}(\ell)] \tag{15-9}$$

where

$\bar{y}_s^*(\ell)$ is the synthetic estimate of the criterion variable for the ℓ-th PSU derived from procedure s,

$\bar{y}(\ell)$ is an unbiased estimator of the criterion variable for the ℓ-th PSU, and

$\text{Var}[\bar{y}(\ell)]$ is an unbiased estimate of the variance of $\bar{y}(\ell)$.

Consequently, the average mean square error of synthetic estimator s over all 59 RTI PSUs can be estimated as:

$$\text{AMSE}[\bar{y}_s^*] = \sum_{\ell=1}^{59} \text{MSE}\,[\bar{y}_s^*(\ell)] / 59 \tag{15-10}$$

Table 15-2. Average Mean Square Errors of Alternative NMCES Synthetic Estimates of Health Insurance Coverage

	Average Mean Square Error			
Criterion Variable	Post-Stratified Estimator	Sample-Regression Estimator	Composite Estimator $C=\frac{1}{2}$	Composite Estimator $C=1/[1+R(\ell)]$
Percent Medicaid	40.17	35.21	35.80	31.98
Percent Medicare	17.91	13.24	14.48	10.36
Percent Privately Insured	124.77	106.69	111.16	96.30
Percent Never Insured	64.22	47.27	51.52	39.87

These average mean square errors are presented in Table 15-2 for the four variables of interest and the alternative synthetic estimators.

Relative root mean square errors for the alternative synthetic estimators were approximated to serve as an additional measure of accuracy. The estimated relative root mean square error takes the form:

$$\mathrm{RRMSE}[\bar{y}_s^*(\ell)] = \{\mathrm{MSE}[\bar{y}_s^*(\ell)]\}^{\frac{1}{2}} \;/\; \bar{y}_s^*(\ell) \qquad (15\text{-}11)$$

Similarly, the average relative root mean square error of synthetic estimator s over all 59 RTI PSUs can be estimated as:

$$\mathrm{ARRMSE}[\bar{y}_s^*] = \sum_{\ell=1}^{59} \mathrm{RRMSE}[\bar{y}_s^*(\ell)] \;/\; 59 \qquad (15\text{-}12)$$

Comparisons of average relative root mean square errors between alternative synthetic estimators may be made by examining Table 15-3.

Table 15-3. Average Relative Root Mean Square Error for the Alternative Synthetic Estimates of Health Insurance Coverage

Criterion Variable	Average Relative Root Mean Square Error			
	Post-Stratified Estimator	Sample-Regression Estimator	Composite Estimator $C=\frac{1}{2}$	Composite Estimator $C=1/[1+R(\ell$
Percent Medicaid	0.6194	0.6361	0.6130	0.5932
Percent Medicare	0.3686	0.3076	0.3264	0.2743
Percent Privately Insured	0.1491	0.1382	0.1409	0.1307
Percent Never Insured	0.3235	0.2759	0.2889	0.2567

all insurance indicators. Although the correlation coefficients (R^2) for the regression models were only moderate, there was no indication of model "lack-of-fit" for the assumed linear relationship. In addition, the relatively small number of NORC PSUs available for estimation of post-strata means for the reliability study reduced the potential number of post-strata in the model. Use of the entire 135 NMCES PSUs in post-strata formation would improve the discrimination and predictive capacity of the post-stratified estimator. However, the sample-regression approach might still have proven more reliable.

As anticipated, the composite estimator whose weight was selected to minimize the mean square error was consistently more accurate than the other synthetic estimators. However, application of this strategy requires <u>a priori</u> estimates of the mean square errors of the component estimators. Most applied settings for synthetic estimation may not have a mechanism for satisfying this requirement. The alternative composite estimator ($C = \frac{1}{2}$) was less accurate than the sample-regression estimator but more accurate than the post-stratified estimator. This method always allows one the advantage of selecting an estimator that is more accurate than the component estimator with the largest mean square error.

Appropriate assumptions regarding the mean square errors of the component estimators, even when direct estimates are unavailable, increase the likelihood of weight selections that approximate the optimal weights. Since composite estimators are rather insensitive to variation in the weights, weights derived from mean square error estimates of related studies may yield a composite estimator more accurate than either of its component estimators (Schaible, 1979). When good judgment is possible with respect to the choice of component estimators and weights, the NMCES study for the health insurance coverage measures indicates the composite estimator is the recommended synthetic estimation strategy for adoption among all appropriate alternative techniques. The method considered in this chapter could also be used to determine the appropriate synthetic estimator for other types of criterion variables being estimated using NMCES data.

15.6 Summary

Reliable estimates of parameters at local and state levels are difficult, if not impossible, to obtain directly from sample surveys, primarily due to cost constraints imposed by the sample size that would be required. The need for such estimates has served as the motivating force in the development of alternative procedures to obtain reliable small area statistics. The selection of the optimal synthetic estimation strategy is a function of the types of symptomatic information that are available, the appropriateness of a specified prediction model, and the number of available sample observational units (i.e., PSUs). Once appropriate estimation methods have been determined, use of a composite estimator with judiciously chosen weights can prevent the selection of the less accurate technique and may potentially yield an estimate of superior precision.

References

Anderberg, M. R. (1973). Cluster Analysis for Applications, New York: Academic Press.

Bureau of the Census (1977). County and City Data Book 1977, U.S. Department of Commerce, Washington, DC: U.S. Government Printing Office.

Cochran, W. G. (1963). Sampling Techniques, New York: John Wiley and Sons.

Cohen, S. B. (1979). A Modified Approach to Small Area Estimation. Synthetic Estimates for Small Areas, National Institute on Drug Abuse Research Monograph 24, Washington, DC: U.S. Government Printing Office, 98-134.

Cohen, S. B. (1980). A Comparative Study of Alternative Synthetic Estimation Strategies With Applications to Data from the National Health Care Expenditure Study. Proceedings of the American Statistical Association, Survey Research Methods Section, 595-600.

Cohen, S. B. and W. D. Kalsbeek (1977). An Alternative Strategy for Estimating the Parameters of Local Areas. Proceedings of the American Statistical Association, Social Statistics Section, 781-786.

Dalenius, T. and Hodges, J. L. (1959). Minimum Variance Stratification. Journal of the American Statistical Association, 54, 88-101.

Ericksen, E. P. (1974). A Regression Method for Estimating Population Changes of Local Areas. Journal of the American Statistical Association, 69, 867-875.

Everitt, B. S. (1980). Cluster Analysis, Second Edition, London: Heineman Educational Books Ltd.

Gonzalez, M. E. (1973). Use and Evaluation of Synthetic Estimates. Proceedings of the American Statistical Association, Social Statistics Section, 33-36.

Gonzalez, M. E. and J. Waksberg (1973). Estimation of the Error of Synthetic Estimates. Presented at the first meeting of the International Association of Survey Statisticians, Vienna, Austria. Available from the U.S. Bureau of the Census.

Johnson, S. C. (1967). Hierarchical Clustering Schemes. Psychometrika, 32, 241-254.

Kalsbeek, W. D. (1973). A Method for Obtaining Local Postcensal Estimates for Several Types of Variables, Ph.D. dissertation, University of Michigan, Ann Arbor, Michigan.

National Center for Health Statistics (1968). Synthetic Estimates of Disability, PHS Publication No. 1759, Washington, DC: U.S. Government Printing Office.

Schaible, W. L. (1979). A Composite Estimator for Small Area Statistics. Synthetic Estimates for Small Areas, National Institute on Drug Abuse Research Monograph 24, Washington, DC: U.S. Government Printing Office, 36-53.

Schaible, W. L., D. B. Brock, and G. A. Schnack (1977). An Empirical Comparison of the Simple Inflation, Synthetic and Composite Estimators for Small Areas. Proceedings of the American Statistical Association, Social Statistics Section, 1017-1021.

Index